Impacts of Shallow Geothermal Energy on Groundwater Quality

KWR Watercycle Research Institute Series

Impacts of Shallow Geothermal Energy on Groundwater Quality

Matthijs Bonte

IWA PUBLISHING

Published by **IWA Publishing**
Alliance House
12 Caxton Street
London SW1H 0QS, UK
Telephone: +44 (0)20 7654 5500
Fax: +44 (0)20 7654 5555
Email: publications@iwap.co.uk
Web: www.iwapublishing.com

First published 2015

British Library Cataloguing in Publication Data
A CIP catalogue record for this book is available from the British Library

ISBN 9781780406817 (Paperback)
ISBN 9781780406824 (eBook)

KWR Watercycle Research Institute
"Bridging science to practice" – that is the mission of KWR Watercycle Research Institute. Our shareholders are the ten Dutch water companies. KWR researchers work at the interface of science, business and society. Their strength lies in their ability to translate scientific knowledge into practical and implementable solutions for end-users in the Dutch and European water sectors. KWR has developed a solid reputation as a top-level innovation accelerator and international network builder.

Contents

Chapter 1

Introduction

1.1 BACKGROUND

Shallow geothermal energy (SGE) systems use the subsurface as a source for thermal energy or as a medium to store seasonal heat. These systems provide heating and cooling in the built environment, and are considered to be a technology that can contribute to the reduction of greenhouse gas emissions (Blum *et al.* 2010; Bayer *et al.* 2012). The reduction of greenhouse gas emission by SGE systems strongly depends on the source of electricity used in the system. The potential CO_2 emission reduction at full market penetration is estimated by Bayer *et al.* (2012) to be around 30% compared to conventional heating and cooling systems, for a selection of 19 EU countries for the current (mainly fossil fuelled) electricity mix. This figure is expected to increase in the future with both the fraction of renewables in the electricity mix and the efficiency of heat pumps increasing. Generally, two types of SGE systems are discerned: open and closed systems (Sanner *et al.* 2003; Mustafa Omer, 2008; Hähnlein *et al.* 2013).

Open systems, also called aquifer thermal energy storage (ATES) or open loop ground source heat pumps, use groundwater to store or harvest heat (Meyer & Todd, 1973). These systems generally operate with one or more sets of extraction and injection wells to circulate groundwater. During winter, groundwater is extracted and used in combination with a heat exchanger and heat pump to provide heating (Figure 1.1a). Heat is extracted from groundwater using a heat exchanger, thereby decreasing the temperature of the extracted groundwater, typically from an ambient temperature of 9–12°C to 5–10°C (under Dutch conditions). The chilled groundwater is re-injected back into the aquifer, and a bubble of relatively cold water is created. During summer, the system flows reverse aiming to recover the relatively cold groundwater and use it to cool buildings (Figure 1.1b). This raises the groundwater temperature to a value typically between 15°C and 20°C, before it is again re-injected into the aquifer creating a warm bubble of groundwater.

The maximum allowable temperature of re-injected groundwater in the Netherlands is 25°C, which is similar to many other European countries (Hähnlein *et al.* 2010a). When heating buildings with ATES, a heat pump is generally used to increase the temperature of the building circulation fluid to a sufficiently high level for space heating. Cooling is generally directly possible, without use of heat pumps which makes these systems especially energy efficient for cooling purposes (Vanhoudt *et al.* 2011). Open systems have been sporadically used for heat storage at much higher temperatures, up to 100°C especially at the early years of the technology (Holm *et al.* 1987; Perlinger *et al.* 1987; Brons *et al.* 1991; Griffioen & Appelo, 1993). Recently, interest in the Netherlands for higher temperature storage is increasing, for example in the horticultural sector where high temperature heat is available for storage.

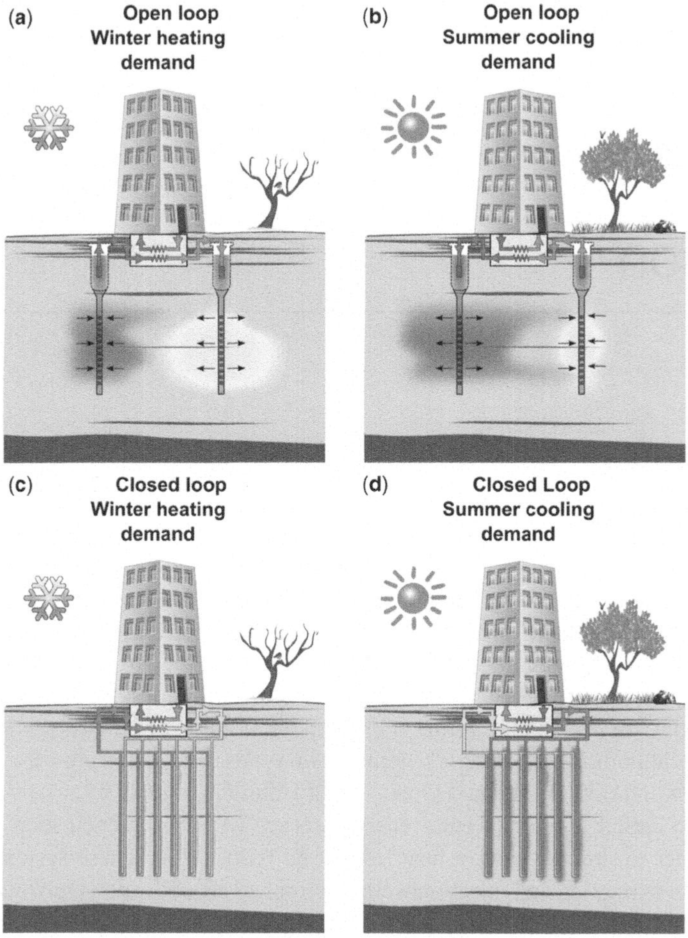

Figure 1.1 Different types of SGE systems: (a) Open loop ground source heat pump in winter (also known as aquifer thermal energy storage, ATES). (b) Open loop ground source heat pump in summer. (c) Closed loop ground source heat pump in winter (also known as borehole thermal energy storage, BTES). (d) Closed loop ground source heat pump in summer.

The second type of system, closed systems, also called borehole thermal energy systems (BTES), or closed loop ground source heat pumps use plastic conductor pipes and one or more boreholes (vertical closed systems, shown in Figure 1.1c & d), trenches (horizontal closed systems), baskets, or even plastic pipes within foundation piles (Mustafa Omer, 2008). Water, often mixed with an anti-freeze fluid, is circulated through these pipes and used to extract heat from the underground. The main difference with the open systems is that groundwater is not extracted during the process. Closed systems generally require more boreholes to deliver the same heating power compared to open systems. Large BTES systems generally require a reasonable balance between heating and cooling load to the subsurface to prevent a subsurface temperature buildup (Ramamoorthy *et al.* 2001).

1.2 SHALLOW GEOTHERMAL ENERGY, SUBSURFACE USE AND DRINKING WATER PRODUCTION

SGE is in many cases a financially viable and an energy efficient way of acclimatizing houses and buildings. For this reason, Dutch and many other European governments are actively stimulating the application of shallow geothermal energy which has led to a strong growth in the application of these systems, both in the Netherlands, and world-wide.

Dutch provincial groundwater data show that the number of ATES licenses increased from 29 systems in 1995 to around 1800 systems in the Netherlands in 2012 (Figure 1.2a). These figures do not include small ATES systems (used for small utility buildings) with a capacity of <10 m^3 $hour^{-1}$ for which no license is required. Data for BTES systems are not available because no license is required for their construction. The Dutch Bureau of Statistics (CBS, 2008) estimated an increase from 24 in 1996 to around 23.000 BTES boreholes in 2007 (Figure 1.2b). Note that one BTES system generally consists of multiple boreholes. Similar strong growth rates are reported in European countries like Switzerland, Sweden, and Germany (Sanner *et al.* 2003), China (Gao *et al.* 2009), and the US (Lund & Bertani, 2010).

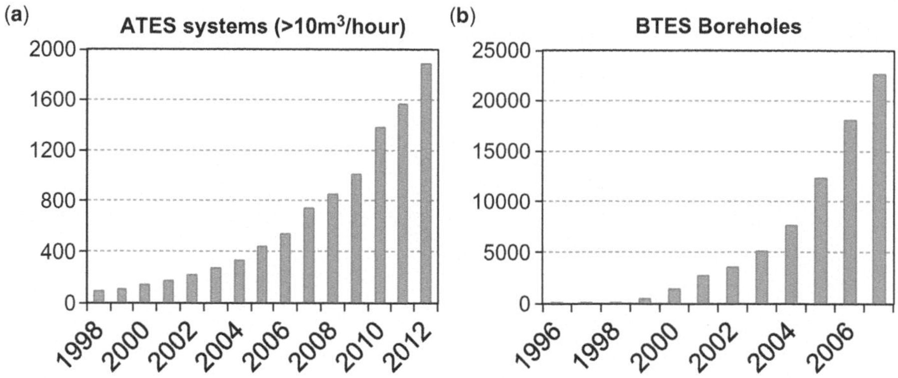

Figure 1.2 (a) Development of aquifer thermal energy storage (ATES) systems. (b) borehole thermal energy system (BTES) systems in the Netherlands. ATES data were sourced from provincial authorities; BTES data were from CBS (2008).

Groundwater licensing data show that 1100×10^6 m^3 $year^{-1}$ is allocated for use in ATES systems in the Netherlands as of 2012 (data from provincial groundwater database). However, actual usage data obtained from the provincial groundwater databases for a selected number of sites showed that ATES systems use, on average, 50% of their licensed volume. Therefore, the actual volume of groundwater circulated is currently estimated to be around 550×10^6 m^3 $year^{-1}$. This water volume is considerable compared to the total annual groundwater extraction in the Netherlands of around 1700×10^6 m^3 (De Vries, 2007). The latter is composed of 1000 million m^3 used for drinking water (of which around 18% originates from artificial recharge), 300×10^6 m^3 for industrial water use and 300–400 million m^3 for agricultural use. The total annual groundwater recharge in the Netherlands is estimated to be around 9000×10^6 m^3 (Dufour, 2000). Given the growth rate of ATES, it is expected that ATES will be the largest groundwater user before 2020. Key difference between ATES and these other uses is that with ATES the groundwater is re-injected, while other users do not return the water to the aquifer. BTES systems harvest or discharge heat to the subsurface and do not extract groundwater.

Despite that SGE systems have no net groundwater extraction, the physical, chemical and microbiological status of groundwater may be impacted upon. In 2009, the Dutch ministerial Technical Committee on Soil (TCB, 2009) reviewed the impacts by SGE and concluded that the relatively low and constant temperature is an valuable asset and that mixing of different water qualities and temperature changes may impact on the natural buffering capacity of the subsurface. The TCB also notes that the long term impacts of SGE on the subsurface are unclear and therefore recommend to maintain the maximum allowed year average re-injection temperature of 25°C (with incidental peak of 30°C) and restrictions for SGE in groundwater protection zones.

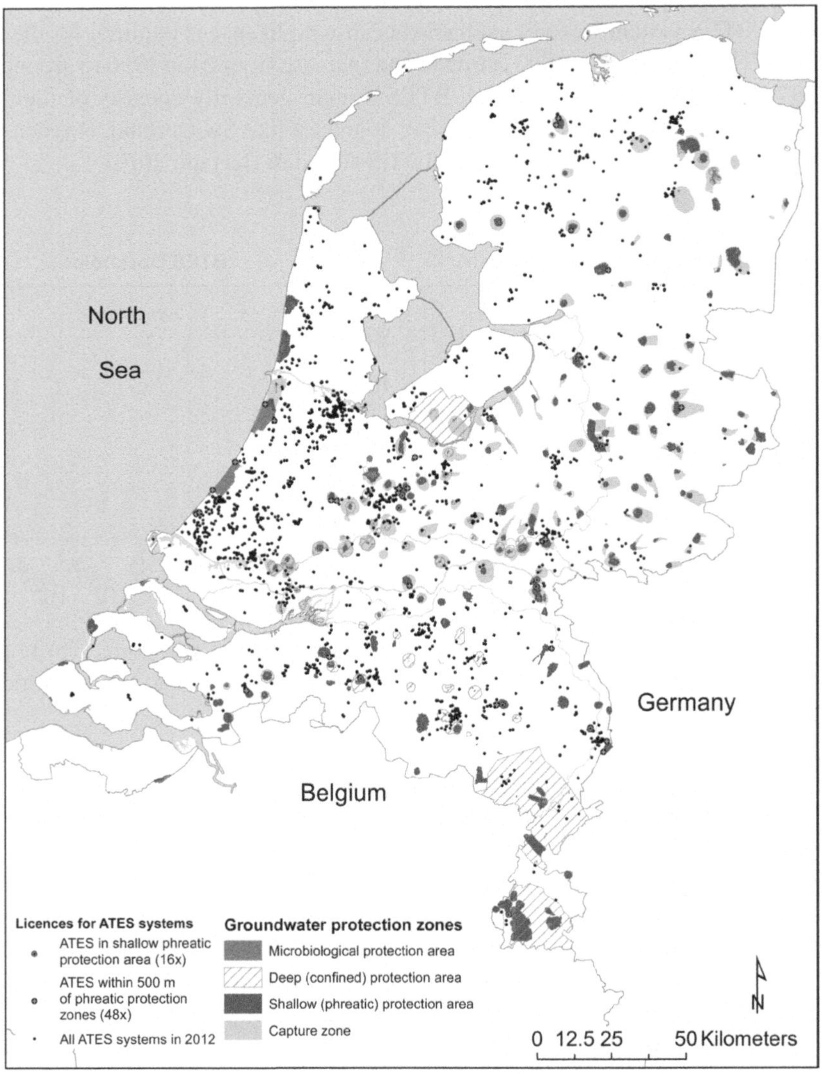

Figure 1.3 Map showing the locations of aquifer thermal energy storage (ATES) systems and groundwater protection zones for public supply well fields in the Netherlands. ATES data were obtained from provincial authorities; data on groundwater protection zones are courtesy of the National Institute of Public Health.

The strong growth and corresponding increasing volumetric water claim have led to an increasing number of sites where ATES systems are planned or were built near public water supply well fields (Figure 1.3). In 2012, 17 ATES systems were located in a groundwater protection area and an additional 52 systems were located within 500 m distance from a protection zone. This growth of subsurface use by SGE is added to a trend where many other functions are going underground: infrastructure (tunnels, parking), water storage (aquifer storage and recovery, storm water infiltration), et cetera. While up to a decade ago, drinking water companies were the main subsurface user in the Netherlands, now many sectors have interests in the subsurface space. This resulted in transformation in the objectives of soil policy from mainly soil and groundwater quality protection towards stimulating the sustainable utilization of the underground while protecting groundwater resources. In order to find a balance between subsurface utilization and groundwater protection, knowledge is required on the impacts of subsurface uses, and in particular those of SGE. Without clear and scientifically underpinned policy, potentially unwanted risks might be taken at vulnerable locations such as near well fields used for drinking water production, while at other sites the application of SGE is avoided without proper reasons.

In order to fill this knowledge gap, the water sector commissioned a research project (through the Joint Research Program, BTO) on the impacts of SGE on groundwater quality and drinking water production which resulted in this book. Related to this project is the PhD project of Phillip Visser (funded through TTIW/Wetsus) which focuses on the thermal impacts of SGE in relation to other stresses to temperature regime such as climate change and land use changes.

It is worthwhile mentioning that these are not the only research projects looking at the impacts of SGE carried out in this time frame. At the TU Delft, smart grids are being developed for ATES by Martin Bloemendal to optimize the use of the subsurface. Two other research projects carried out at Wageningen University by Zhuobiao Ni and Wijb Sommer, deal with the interaction of groundwater contaminants and ATES systems. The project 'Meer met Bodemenergie' investigated a large number of SGE in both clean and contaminated groundwater systems. Together, these research programs aim to form a sound basis to regulate use SGE in various groundwater environments.

1.3 RESEARCH OBJECTIVE AND QUESTIONS

The main objective of this book is to obtain a quantitative understanding of the impacts of shallow geothermal energy on groundwater quality in relation to drinking water production and other uses. In order to reach this objective, I aim to answer to following questions:

- What is reported in literature on the effects of SGE on groundwater quality and is current groundwater protection policy adequately addressing these aspects?
- Does an ATES system near a drinking water station impact on groundwater quality, and if so, what are the driving physical and chemical processes for this change?
- What are the impacts of changing temperatures on groundwater quality? In particular regarding:
    - the mobility of naturally occurring trace elements and heavy metals;
    - the rate and occurrence of biochemical (redox) processes;
    - the aquifer's microbial community and its relation to the biochemical processes.
- What are the key geochemical processes determining the impacts of SGE on groundwater quality, can these factors be incorporated in a predictive hydrochemical model to describe impacts from circulating groundwater with varying temperatures in an ATES system.

These sub-questions will be answered in the following chapters, the main research question will be discussed in the last chapter (summary and synthesis) where I will compile the insights gathered from this research and place them in the context of drinking water production.

1.4 METHODOLOGY AND OUTLINE

The outline of this work is structured as follows:

- In Chapter 2, an overview of risks posed by SGE systems on groundwater quality is presented based on an overview of previous literature on SGE in relation to groundwater quality. Existing policy and licensing arrangements for SGE in the Netherlands and a selection of European countries are described as well as the capability of the Dutch policy on SGE to minimize or mitigate risks from SGE on groundwater resources. The increasing pressure on the subsurface is discussed and placed in the context of subsurface policy reform.
- In Chapter 3, I present a water quality data set obtained from an ATES system operating between 7 and 16°C located 500 m distance away from a large public water supply well field. The data were analyzed using a groundwater flow and solute transport model (MODFLOW/MT3D) which was used to elucidate the dominant processes controlling observed water quality changes.
- In Chapter 4, laboratory column experiments were used to evaluate temperature effects on leaching behavior of trace elements and heavy metals. The experiments were carried out at 5, 11, 25 and 60°C applying a 1 day residence time in order to provide information on SGE impacts at the currently allowed temperature range and more extreme high temperature.
- In Chapter 5, the column setup is used applying longer residence times focusing on the occurrence and rates of redox processes (iron-reduction, sulfate-reduction, methanogenesis) over a wide temperature range of 5–80°C. A combination of microbiological techniques (T-RFLP and 454 pyrosequencing of PCR-amplified bacterial 16S rRNA genes, and DGGE of PCR fragments of archaeal 16S rRNA genes) was used to investigate changes in the microbial community.
- In Chapter 6, the experimental data gathered using the experiments described in Chapter 4 were integrated and simulated using a reactive transport model (based on the PHREEQC code) to improve the conceptual model of temperature impacts on groundwater quality. The model was subsequently applied to several hypothetical ATES system using a coupled axi-symmetrical flow tube setup.
- In the last chapter, a summary is provided followed by a synthesis and discussion of the key results. This is done by first discussing what the observed impacts on groundwater quality imply for drinking water treatment and groundwater monitoring, followed by a reflection on the current SGE and groundwater protection policy. Lastly, an overview of new research questions resulting from this work is given.

Chapter 2

Shallow geothermal energy: A review of impacts on groundwater quality and policy in the Netherlands and European Union

ABSTRACT

In this chapter we present an overview of the risks shallow geothermal energy (SGE) can impose on the groundwater system, drinking water production, and the subsurface environment in general. Existing policy and licensing arrangements for SGE in the Netherlands are described as well as the capability of the current (and future) Dutch policy and legal framework to minimize or mitigate risks from SGE on groundwater resources. A survey at EU Member State level indicated that regulation and research on the potential impacts of SGE on groundwater resources were lagging behind the technological development of, and ever-growing demand for, this renewable energy source. We have recognized three main issues that should be addressed to secure sustainable application of SGE: (1) Scientific research is required to further elucidate the impacts of SGE on groundwater; (2) Cross-sectoral subsurface planning is required to minimize negative conflicts between SGE and other subsurface interests; (3) EU wide guidelines and standards are required for quality assurance and control when installing a SGE system.

2.1 INTRODUCTION

As shown in Chapter 1, SGE technologies, including Aquifer Thermal Energy Storage (ATES) and Borehole Thermal Energy Systems (BTES) are seen as promising techniques that can contribute significantly to reduce the emission of greenhouse gasses. This has led to a strong growth in the deployment of these technologies, often in aquifers used for drinking water production, and as a consequence, regulators, authorities, and drinking water companies are asking questions on the impacts of SGE on groundwater conditions. Many authorities in Europe are in the process of reforming their groundwater policy that traditionally dealt primarily with protecting groundwater resources for drinking water production towards balancing groundwater protection with subsurface utilization.

The main question in this chapter is whether Dutch and other EU national governments are ready to regulate the increasing use of SGE while minimizing potentially adverse effects on the underground environment. We will (i) detail the risks that SGE may impose on groundwater quality and drinking water

*[*]Modified and updated with comtemporary references from Bonte M., Stuyfzand P. J., Hulsmann A. and Van Beelen P. (2011) Underground thermal energy storage: environmental risks and policy developments in the Netherlands and European Union. Ecology and Society, **16(1)**, 22.*

production; (ii) describe current and future policy and licensing arrangements for SGE in the Netherlands and the EU; and (iii) discuss the need for cross-sectoral subsurface spatial planning in order to minimize negative impacts from different subsurface activities. These three aspects will result in recommendations for EU member states currently in the process of shaping their policy on groundwater protection and SGE.

2.2 A REVIEW ON THE POTENTIAL IMPACTS OF SGE ON GROUNDWATER QUALITY

In order to assess the risks from SGE systems to groundwater quality and ecology, and drinking water production, we reviewed common practice in SGE operation in relation to drinking water production (Bonte *et al.* 2008; Stuyfzand *et al.* 2008). From this review, four groups of environmental risks arising from SGE can be distinguished: (1) hydrological impacts; (2) thermal impacts (3) chemical impacts; and (4) microbiological impacts. Table 2.1 presents a summary of the different effects, the probability of occurrence, the consequences if an impact occurs, and the associated risk (probability multiplied by its consequence). This table is based predominantly on expert opinion (Stuyfzand *et al.* 2008) and guided further research on the effects of SGE. Some effects have a small chance but potentially far-reaching consequence (for example leaching and leaking of heavy metals or antifreeze media) and therefore have a high risk associated with them. Other effects are certain to occur but the consequences are limited.

Table 2.1 Qualitative overview of risks of low-temperature (<30°C) underground thermal energy storage on groundwater systems. Based on (Stuyfzand *et al.* 2008).

Effects of SGE	Probability*	Consequence†	Risk‡
Hydrological impacts			
Changing water levels and fluxes	++	Desiccation, water logging, settlements	±
Changing of other well's capture zone	++	Increasing vulnerability, pollutions	++
Poorly sealed boreholes	+	Cross aquifer flow	++
Thermal impacts			
Changing water temperature	++	Temperature, reaction kinetics	+
Chemical impacts			
Mixing processes	++	Salinity, IMIPO, OMIPO	++
Mobilization of groundwater pollution plumes	±	IMIPO, OMIPO	++
Oxidation of organic matter	±	Nutrients, DOC, color	+
Oxidation of iron-sulfides	±	Fe, SO_4, As, Ni, Co, Zn	+
Dissolution/precipitation carbonates	±	Ca, HCO_3, Sr	±
Dissolution/precipitation of silicates	±	SiO_2	±
Leaching from installation materials	±	Cd, Cu, Cr, Ni, Pb, VC	+
Leaking anti-freeze fluids or additives	±	Glycol, biocides & corrosion inhibitors	++
Microbiological impacts			
Introduction of pathogens	–	Pathogens	±
Changing biodegradation rate	±	Nutrients, IMIPO, OMIPO	– or +
Changing microbiological population	+	Unknown	?

*– small probability, ± medium probability, + large probability, ++ will always occur;
†IMIPO: Inorganic micro pollutants; OMIPO, Organic micro pollutants; DOC, Dissolved Organic Carbon; VC, Vinyl chloride.
‡– negative risk (opportunity), ± no apparent risk, + small risk, ++ high risk.

2.2.1 Hydrological impacts

The hydrological risks relate to groundwater quantity aspects and include groundwater levels, flow velocities, and the capture zone of public water supply well fields. Although ATES systems have no net extraction as groundwater is re-injected in the same aquifer, perturbations in the groundwater flow pattern can be noticeable to a distance of up to several kilometers (Ferguson, 2006). This situation can have a direct impact on the size and location of the capture zone of a groundwater abstraction well field and is explained conceptually in Figure 2.1. Situation A in this figure shows a typical groundwater extraction well with its capture zone and the legally enforced groundwater protection zone (usually the area where the travel time to the well is less than 25 years). Situation B shows the impact that an ATES system may have on the groundwater flow paths towards the extraction well. The alternating extractions and injections will cause the flow paths to become jagged (in Figure 2.1 this process is strongly exaggerated for visualization purposes). The consequence of this is yet unclear; the effective residence time of water is not expected to be impacted upon but groundwater will have a longer flow path and higher velocity in its way to the extraction well. In situation C, one well of the ATES is within the capture zone of the extraction well. In this case, the ATES system periodically takes groundwater from outside the original capture zone of the extraction well and injects it inside the capture zone. This implies that the capture zone of the extraction well is enlarged with the capture zone of the ATES system. The latter process may also occur for situation B. The actual enlargement in relation to the capture zone will depend amongst others on the volume of water circulated in the ATES system, the thickness of the aquifer and the annual extraction volume.

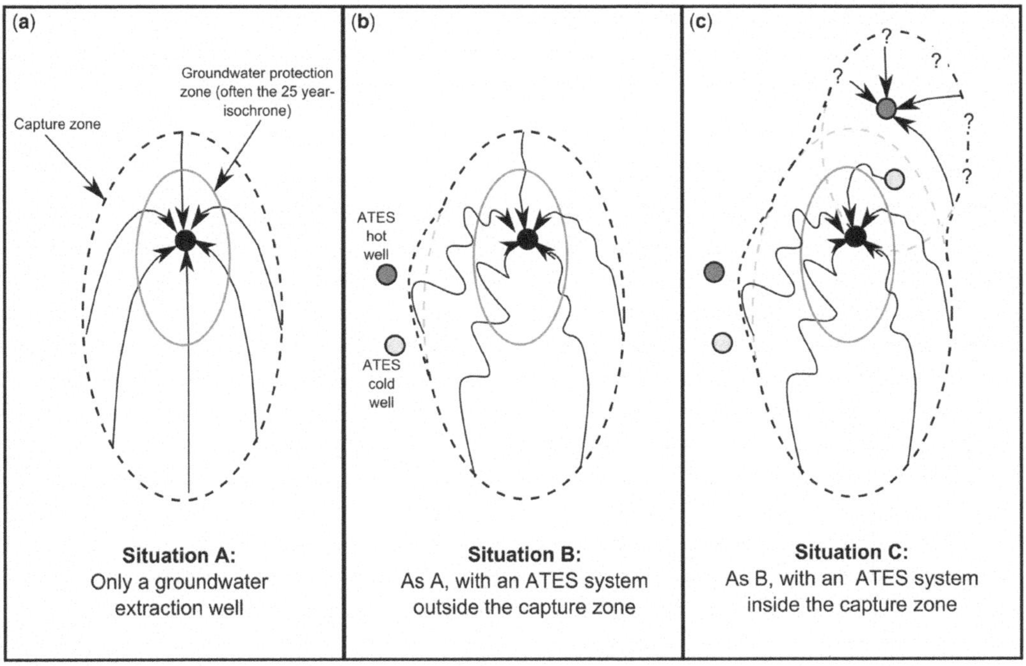

Figure 2.1 Conceptual illustration of the effects of an aquifer thermal energy storage (ATES) system: alone (a), just outside the capture zone of a public supply well (b), and just inside the capture zone of a public supply well (c). Arrows indicate the path of groundwater flow; dark grey and light grey circles are the warm and cold wells, respectively, of an ATES system.

Depending on aquifer conditions like heterogeneity, sediment reactivity, and land use, this may cause quality changes at the groundwater extraction well, in positive or negative sense. If surface waters are present within the capture zone, ATES may alter the nature of groundwater-surface water interactions. For example, a normally gaining stream (receiving groundwater) may change into an alternating losing and gaining stream depending on ATES operation. It should be noted that while capture zones in spatial planning are often considered a steady state feature, in reality the location will vary due to the transient nature of groundwater recharge and surface water-groundwater interactions (Rock & Kupfersberger, 2002). The effect of transient pumping at an ATES system can act cumulatively and exacerbate these transient variations.

Groundwater flow patterns may also be influenced if ATES or BTES boreholes are constructed improperly. Boreholes that lack adequate clay (frequently bentonite) or grout plugs to separate aquifers or boreholes screened in several aquifers may provide preferential flow paths for contaminants or cross-aquifer flow (Avci, 1992; Lacombe *et al.* 1995; Chesnaux *et al.* 2006; Santi *et al.* 2006; Mayo, 2010). Minimizing drilling costs may result in selecting a drilling method that does not allow precise geological logging required to determine the depth of aquitards to be plugged. In the Netherlands, BTES boreholes constructed before July 2013 did not require a license nor registration and although a protocol exists for drilling these boreholes which describes how aquitards are to be plugged, the lack of licensing and regulatory enforcement makes it questionable whether contractors have always followed these rules. As of July 2013, regulations are in place, but it is not yet known whether municipalities will always enforce that rules will be followed. In some countries, regulators enforce the entire annulus of BTES boreholes to be grouted for increasing the thermal efficiency of the well and reducing the risk of cross contamination. Although cementing may initially limit vertical flow in the borehole, de-bonding of conductor pipe and grout due to differences in thermal expansion behavior may still result in preferential flow paths later on (Philippacopoulos & Berndt, 2001). Enhanced grout types, containing a super plasticizer and bentonite, can reduce the likelihood of de-bonding (Philippacopoulos & Berndt, 2001).

2.2.2 Thermal impacts

An inventory of 67 ATES systems showed that almost none of the investigated systems had a thermal balance (IF Technology, 2007), meaning that heat is discharged into, or harvested from the aquifer and a long-term warming or cooling of groundwater is occurring. A study in Winnipeg, Canada, showed that using the aquifer solely for cooling purposes is not sustainable due to long-term rising groundwater temperatures which can cause short circuiting of warm and cold wells and a reduction in cooling capacity (Ferguson & Woodbury, 2006). Apart from reduced efficiency of SGE due to changing temperatures, downstream uses of groundwater and aqueous ecosystems can be negatively impacted (Ferguson, 2009). The downstream impacts were investigated (Freedman *et al.* 2012) for an open loop cooling system around 2 km upstream from the Columbia River. Freedman *et al.* (2012) concluded that at the operating regime the system did not significantly impact upon the natural temperature regime of the river. Freedman *et al.* (2012) also noted that there was a conflict between operational efficiency and downstream impacts as optimal operation demanded high circulation rates with a relatively small temperature drop (facilitating an efficient heat discharge) causing a steep gradient to the river and higher impacts at the river.

A study on closed loop systems by Hähnlein *et al.* (2010b) showed that these systems can operate safely without negatively impacting each other when the spacing between them is at a minimum distance of 10 m. The diverging results of these three studies are likely due to differences in heat demand from the aquifer, type of system (open versus closed) and aquifer conditions (groundwater flow velocity, aquifer thickness, porosity). For large scale systems in an area where other uses of groundwater may be negatively impacted upon, a hydrological impact assessment should thus be made to prevent negative interference between systems. Hecht-Mendez *et al.* (2010) showed that temperature driven buoyancy effects for typical temperature ranges ($\Delta T < 10°C$) can generally be neglected allowing the use of the solute transport model MT3D-MS for thermal modeling.

In order to adequately assess the long term cumulative effects of heat discharge, the autonomous trends due to changing environmental stresses to the groundwater system should also be considered. For example, when analyzing the thermal impacts of SGE on the underground, the temperature effects of climate change and urbanization (urban heat island effect) on the aquifer system should be accounted for (Taniguchi & Uemura, 2005; Ferguson & Woodbury, 2007; Kooi, 2008). The key question is whether the effects of SGE can be neglected compared to these stresses, or should SGE be considered as yet another thermal stress on a system that is already threatened.

2.2.3 Chemical impacts in clean groundwater systems

ATES systems may influence groundwater chemistry by (i) the mixing of different types of groundwater (for example aerobic and anoxic), or (ii) via changes to the thermal regime of the water impacting on sediment-groundwater interaction.

In water table aquifers, the quality of groundwater varies with depth due to chemical reactions of infiltrating rainwater with reactive soil and aquifer compounds such as carbonates, pyrite, and organic carbon (Stuyfzand, 1999; Appelo & Postma, 2005). This results in a vertical groundwater quality gradient with oxidized, nitrate-rich shallow groundwater and reduced iron-rich deeper groundwater. The aquifer acts as a groundwater quality buffer and assures a stable and relatively high water quality, when compared to surface water. In regions where fresh groundwater overlies saline groundwater (which is particularly the case in the west of the Netherlands and in many other deltas), an improper well screen setting of an ATES system can cause salinization of fresh groundwater (Oostrum *et al.* 2008). In both cases (redox or salinity stratification), the ATES system extracts groundwater from a certain depth interval, mixes it, and re-injects the mixture into the aquifer, and depending on the screen setting in relation to vertical quality gradients, the system levels out the natural vertical quality gradient. This aspect was earlier discussed by (TCB, 2009) and is further investigated in Chapter 3. This can have adverse effects for the quality of the groundwater resource, introduce environmental pollutants to greater depths in the aquifer, and minimize the volume of clean uncontaminated groundwater, ultimately increasing the vulnerability of PWSFs or extraction wells for other purposes.

The mixing of different water qualities has recently also been described as the dominating process controlling water quality changes in seven ATES systems in the Netherlands for the 'Meer met bodemenergie' project (Dinkla *et al.* 2012). Besides the effects of mixing, this study also identified elevated concentrations of nickel and arsenic at an ATES system near Rosmalen. These observations were explained by mixing and introduction of shallow (sub)oxic groundwater resulting in pyrite oxidation (Dinkla *et al.* 2012). Monitoring at an ATES system at the contaminated location Strijp-S in Eindhoven revealed elevated concentrations of nickel, arsenic, and chromium which increased in both monitoring wells under direct influence of the ATES system, and reference wells. A satisfactory explanation was not possible, and it was recommended that more monitoring data be collected (Dinkla *et al.* 2012).

Most of the research published internationally on chemical impacts focused on the change in mineral solubility, reaction kinetics, and organic matter oxidation (Holm *et al.* 1987; Appelo *et al.* 1990; Brons *et al.* 1991; Griffioen & Appelo, 1993; Hoyer *et al.* 1994; Arning *et al.* 2006b). One of the most relevant studies for drinking water production is by Brons *et al.* (1991) who looked at mobilization of organic carbon and release of CO_2 from sediments under increased temperature. Dissolved organic carbon (DOC) is often related to coloration of water (Rittmann *et al.* 2002) requiring treatment. Furthermore, sedimentary organic carbon contributes to the buffering capacity of sediments (Griffioen *et al.* 2012) and is thus important in aquifers used for the production of drinking water (TCB, 2009). Incubation experiments by Brons *et al.* (1991) showed that aerobic organic carbon mineralization increased over the entire temperature range (4–95°C) with biological processes dominating <30°C and pure chemical oxidation >55°C. At temperatures >40°C humic acids were mobilized which may form complexes with calcium and magnesium and inhibit carbonate precipitation. The mobilization of DOC >25°C was also reported recently in column experiments by (Jesußek *et al.* 2012). The focus of these studies was however mostly on operational aspects and not on the groundwater quality changes at elevated temperatures in relation to groundwater quality for drinking water production. Especially the behavior of trace elements and heavy metals, commonly present in natural sediments, at typical temperatures changes in SGE has received little attention in literature. This aspect is further investigated in Chapters 4 and 6.

The results of most of the above mentioned studies suggest that most inorganic hydrochemical processes will start to play a significant role at temperatures above 30°C. Yet, seasonal temperature changes were shown to (i) strongly accelerate pyrite oxidation in a deep well injection experiment (Stuyfzand, 1998; Prommer & Stuyfzand, 2005), and (ii) be of great influence on mineralization of organic matter at river bank filtration sites (Greskowiak *et al.* 2006; Sharma *et al.* 2012). A shift in redox conditions was also reported by (Jesußek *et al.* 2012) who reported the dominance of nitrate reduction at 10°C and of iron reduction at 25°C and higher. These studies illustrate that especially redox reactions can be influenced already by small temperature changes. The impact of temperature on redox processes in permanently anoxic aquifer sediments, as typically used for SGE in the Netherlands, has received less attention, and this aspect is further investigated in Chapters 5 and 6.

2.2.4 Chemical impacts in contaminated groundwater systems

In the Netherlands, there are many inner city areas where there is a desire to apply SGE, but the aquifer is contaminated by historic activities such as chemical dry cleaning, oil storage tanks and gasworks. The effect of SGE on such groundwater contamination plumes is yet unclear. An increased groundwater temperature may mobilize otherwise immobile contaminants due to increased solubility and reduced sorption (Knauss & Copenhaver, 1995; Ten Hulscher & Cornelissen, 1996; Knauss *et al.* 2000), or may increase contaminant toxicity (Noyes *et al.* 2009). The mixing of different chemical groundwater types, mobilization of nutrients and DOC, and increasing groundwater temperature may impact (both positively and negatively) on biodegradation (Langwaldt & Puhakka, 2000). Zuurbier *et al.* (2013b) showed with reactive transport modeling that the circulation of groundwater in an ATES system can considerably reduce the contaminant load if biodegradation is sediment limited with a constant rate in time and space. If however a dense non-aqueous phase liquid is present, the dissolution of contaminant mass may create a much larger contaminant plume (Zuurbier, 2010).

Field and laboratory experiments showed that reductive dechlorination is not readily stimulated by a temperature increase alone (Friis *et al.* 2006b; Fletcher *et al.* 2011), but requires bio-augmentation composed of a dechlorinating culture and an electron donor (Friis *et al.* 2006a). In the Dutch cities of Eindhoven, Apeldoorn and Utrecht, field and feasibility studies are currently being conducted to combine ATES with a remediation system (Slenders *et al.* 2010), and also laboratory investigations are being carried out on sediments from Utrecht investigating the combination of ATES and remediating chlorinated solvents (Ni *et al.* 2011; Ni *et al.* 2013).

2.2.5 Chemical impacts of BTES systems

BTES systems use cooling fluids that, in the case of a poorly installed, damaged or aged system, could leak into the aquifer. The coolant is often a mixture of water and an antifreeze agent (for example ethylene- or monopropylene glycol), and many pre-mixed circulation fluids contain a biocide and a corrosion inhibitor (Klotzbucher *et al.* 2007). Although both aerobic and anaerobic biodegradation has been observed in soils for frequently used anti-freeze agents such as ethylene glycol, propylene glycol and betaine, the addition of corrosion inhibitors or biocides in BTES systems can inhibit biodegradation (Klotzbucher *et al.* 2007). Typical corrosion inhibitors used in mixtures used for BTES systems are benzotriazole and tolyltriazole (Illieva *et al.* 2012), and also nitrite, boron and molybdenum are typical additives (Schmidt *et al.* 2013). The triazoles not only inhibit the biodegradation of the anti-freeze medium, but they are also relatively toxic to humans and ecosystems, and persistent and mobile in sediments (Illieva *et al.* 2012; Schmidt *et al.* 2013).

The prevalence of leaky BTES systems is currently not known. However, given the growing number of BTES systems in the Netherlands alone, and the lack of regulations to enforce high quality drilling work, the risk of contamination of the groundwater system is likely to increase with time. The latter aspect has improved in the Netherlands with the introduction of new legislation (July 2013) which enforces that BTES systems are to be registered with municipalities and that leak-detection systems are to be used for large systems that have a buffer tank.

2.2.6 Microbiological Impacts

The microbial quality of groundwater is a vulnerable and valuable asset, especially when groundwater is used as a source for drinking water. Groundwater is not a sterile environment but an ecosystem and even anaerobic groundwater contains bacteria (Griebler & Lueders, 2009). In aerobic groundwater, fungi, protozoa, and groundwater invertebrates are also present (Goldscheider *et al.* 2006). Groundwater microorganisms use oxidants such as oxygen, nitrate, and sulfate, and thereby lower the redox potential (Korom, 1992). Especially the removal of nutrients is important for groundwater quality as it reduces the growth of pathogens in groundwater and in drinking water infrastructure.

Since there are many different types of pathogens it is very difficult to prove their overall complete absence in drinking water. Persons have different susceptibilities for pathogens and a minority of the affected human population might become ill for a certain period because of pathogens in drinking water. Standard tests for fecal bacteria in water are effective in detecting fecal pollution but are not proof that pathogens are absent (Winters, 1992).

Microbiological impacts by SGE systems can occur due to changing groundwater temperatures (Hall *et al.* 2008) or the introduction of biologically-available nutrients by well drilling fluids. Previous microbiological research on SGE systems showed that although no evidence is observed for growth or survival of pathogens (Winters, 1992) or increasing cell counts (Schippers & Reichling, 2006), a considerable change in the community composition is shown (Sowers *et al.* 2006; Brielmann *et al.* 2009; Brielmann *et al.* 2011). Brielmann *et al.* (2009) found that the microbial diversity was strongly temperature dependent in groundwater near an SGE site, but also a nearby river had a strong influence. They concluded that although SGE (<18°C) impacts on microbial diversity, the impact is limited compared to seasonal variability and therefore poses no likely threat to drinking water production in shallow aquifers. In a laboratory study, Brielmann *et al.* (2011) confirmed the strong temperature dependence of the microbial diversity over a range of 2–45°C, and they also found that groundwater invertebrates showed little tolerance to temperature changes. Changes in the microbial community can also impact on the SGE system itself: Both Vetter *et al.* (2012) and Lerm *et al.* (2011) report Fe and S oxidizers to be responsible for clogging of the cold storage (5–30°C) well of the German Parliament. Furthermore, (Lerm *et al.* 2013) showed that sulfate reducers may cause corrosion and lead to clogging by precipitating iron sulfides.

2.3 PAST AND CURRENT POLICY FOR SGE

2.3.1 SGE policy in the Netherlands prior to July 2013

Open loop ATES systems with a capacity in excess of 10 m^3 per hour were required to obtain a permit from the provincial government as enforced in the national Water Act. All 12 Dutch provinces have developed their own policy guidelines for permitting ATES systems and no overarching national policy on the implementation of ATES exists. As a result, there are considerable differences in policy across provinces. In general, ATES systems are not allowed within groundwater protection areas for drinking water supply. Provincial authorities can however make an exception (as shown in Figure 1.3).

Policy differences between provinces with regard to ATES systems can be considerable. For example, the province of Zuid-Holland has prohibited the use of the first (or most shallow) aquifer for ATES systems in urban and greenhouse areas. This exclusion aims to minimize conflicts with other subsurface uses such as underground constructions (parking garages) in urban areas and subsurface rainwater storage in greenhouse areas. As a result, ATES systems can only be realized in the deeper (second and third) aquifers (generally >50 m depth below surface), causing small ATES systems often not to be financially viable due to high drilling costs. On the other hand, the province of Noord-Brabant has prohibited ATES systems to be installed deeper than 80 m in an effort to protect the quality of the deep fresh groundwater used for drinking water supply.

Prior to July 2013, closed systems were regulated only in groundwater protection zones under the Dutch national Environmental Protection Act. Groundwater protection zones are set by provinces around locations for groundwater extraction for drinking water production. The provinces have set further environmental regulations in these zones for BTES systems via Provincial Environmental Regulations. In all provinces, drilling or disturbing the soil below two to three meters is prohibited in groundwater protection zones, except for situations where an exemption has been obtained (which is very rare). Outside groundwater protection zones, BTES systems could up to July 2013 be installed without any registration or license.

2.3.2 SGE policy in the Netherlands after July 2013

In 2008, the Dutch Ministry of Housing, Spatial Planning, and the Environment commissioned a group of energy, soil, and water experts to draft a plan to stimulate deployment of SGE whilst considering the potential risks that this technology forms to groundwater and soil quality. Their report (Underground Energy Taskforce, 2009) contains a number of recommendations. The most important being policy changes including: (1) a so-called 'traffic light model' for SGE permits; (2) the introduction of subsurface spatial planning; and (3) a proper distribution of the financial costs and benefits for large-scale SGE development. With the effectuation of the Order in Council (OiC) on SGE (July 2013) some of the recommendations have been implemented.

The first recommendation of the Underground Energy Taskforce (2009) involved a geographical distinction of three SGE application units (the 'traffic light model') for all types of SGE deployment. In green areas SGE can be applied without a permit. In orange areas other subsurface interests exist, such as existing SGE systems, and a permit is required. In red areas other, often more important subsurface interests exist, such as drinking water production, and only under special circumstances a permit is granted. The legal framework to implement this geographical zoning is implemented in the OiC by giving provinces and municipalities the authority to define so called 'interference' regions where subsurface spatial plans can be developed to avoid negative interference between systems. It is noted that a similar approach has successfully been applied in Germany (Eugster & Sanner, 2007) where maps have been drafted clearly showing what type of SGE is allowed where.

The second recommendation involves a more important role for the municipalities in subsurface spatial planning, especially in areas defined as orange and red in the traffic light model. In dense urban areas, such as The Hague or Rotterdam, use of underground space is rapidly increasing. Poorly positioned SGE systems can negatively influence each other causing reduced overall efficiency (Bakr *et al.* 2013). Such issues will need to be addressed by future subsurface planning. In some cities in the Netherlands, subsurface master plans have been developed in which the exploitation of the underground for SGE is regulated. Figure 2.2 shows an example of a subsurface planning map with regions where cold and warm wells of ATES systems can be installed (Witteveen & Bos, 2008) such that the bubbles do not interfere with each other. The orientation of these lanes is to be selected parallel to the direction of groundwater flow. Similar efforts for subsurface master planning dealing with the growing use of the subsurface in general are recently being addressed in literature (Ursej & Kontic, 2007; Bobylev, 2009; Evans *et al.* 2009). Caljé (2010) used Amsterdam as a case study to show that these plans are only suitable for areas with a high heat demand. In areas with sufficient capacity for heat storage in the aquifer, the *laissez faire* (not coordinated) approach is adequate and often more efficient because location of wells can be optimized without restrictions.

Figure 2.2 Subsurface spatial planning map for the second aquifer in the Goudse Poort in Gouda, the Netherlands. Regions indicate with different hatchings indicate areas where warm, cold, and no aquifer thermal energy storage wells can be installed. Modified from (Witteveen & Bos, 2008).

Under the proposed policy changes, the Underground Energy Taskforce expected (in 2009) a growth rate of approximately 30% per year for SGE deployment while this will be around 12% per year without policy changes. These two scenarios yield a total of 18,000 versus 3500 ATES systems in operation in 2020, under changed or current SGE policies, respectively. This in turn results in a CO_2 emission reduction of 0.6 to 2.9 Mton per year for the two scenarios. The total CO_2 emission in the Netherlands is estimated by (CBS, 2008) at 173 Mton in 2007, of which around 40% (70 Mton) is generated in the built environment. No figures are given by the Taskforce on the volume of groundwater used. If we assume a water use per system that is equal to the current situation, the total water use in 2020 will be 1225 to 6300 million m^3 per year depending on the two scenarios. Remembering that the total annual groundwater extraction is 1500 million m^3 and the total annual groundwater recharge amounts to 9000 million m^3, this means that the production of renewable energy is likely to be the largest user of groundwater in the Netherlands in 2020. As discussed in Chapter 1, a key difference with other groundwater users, is that with ATES the groundwater is re-injected, while for drinking water it is not. If we assume an average well screen length of 50 m and a porosity of 0.3, the subsurface space claim for both warm and cold bubbles (thus doubling the volume of water) in 2020 is estimated to be between 163 and 840 km^2 ($2 \times 1225 \times 10^6/(0.3 \times 50 \times 10^6)$ and $2 \times 6300 \times 10^6/(0.3 \times 50 \times 10^6)$). The total area of the Netherlands is 41,543 km^2 of which 3449 km^2 is built-up area (Compendium voor de leefomgeving, 2013), so that the subsurface space claim in built-up areas ranges between 5% and 24%. In fact, the space claim may in some areas be even higher as some distance between adjacent SGE systems is required to avoid interference between the systems, in other areas (for example in Amsterdam) the aquifer is thicker than 50 m and the claim per m^3 of circulated water may be lower.

2.3.3 Policies of European member states on SGE

SGE systems are widespread in Europe, with the longest application history in central Europe (Austria, Germany, and Switzerland) as well as in Sweden. Next, a market developed in the Benelux countries, France, Finland, Ireland, UK, and the eastern European countries. The current strong increase in demand has changed the established market and resulted in the involvement of new players with less experience. Strong quality assurance with accompanied training and certification programs are urgently needed to prevent negative environmental impacts and damage to the public perception of the technology (Eugster & Sanner, 2007).

In 2009, an exploratory inventory was made amongst ENDWARE members (the informal European Network of Drinking Water Regulators). Questions were asked on the penetration level of SGE within their national territory, whether or not the regulators had considered the potential impacts on underground water resources, and the availability of existing or scheduled policy and regulation of underground systems within and outside groundwater protection zones. EU Member States such as Germany, Sweden, the Netherlands, Poland, Estonia, Lithuania, and the United Kingdom recognized the growing issue of energy-related activities in the underground and the potential risk to groundwater supplies designated for the production of drinking water. Estonia, for example, began drafting legislation in 2009 which resulted in legislation as of 2012, which determined that ATES systems require approval under the Water act. In Poland attempts to issue regulations have been made but there was at the time of the 2009 inventory no clear legislation. In the latter country the land-owner has the right to extract the groundwater at their property without any restrictions. Other EU Member States such as Lithuania, Italy, Czech Republic, Portugal, Slovenia, and Cyprus are not aware of any specific requirements or do not know about the technology and related regulations. In Norway the construction of SGE systems is not prohibited but specifications and environmental protection measures are decided upon on a case by case basis.

Comprehensive regulation is in force in traditional SGE countries such as Sweden, the United Kingdom, Switzerland, Germany, and Austria. The CAR (controlled activities regulation) system in Scotland allows the Scottish Environmental Protection Agency (SEPA) to control activities which may have an impact on the water environment (both groundwater and surface water). It is the key legislation that enables Scotland to achieve environmental improvements and to protect and improve the water environment in a sustainable way in line with the aim and the objectives of the Water Framework Directive. Sweden not only uses BTES and ATES systems but also Cavern Thermal Energy Storage (CTES) systems. The regulation for BTES varies between different areas of Sweden and for ATES and CTES the water laws and regulations apply.

Within the countries with a developed market, national or regional water management and/or groundwater protection authorities have published guidelines for the license proceedings and for the construction and operation of the installations. The German VDI (Verein Deutscher Ingenieure VDI-GET) (Reuss & Konstantinidou, 2006) has developed technical guidelines for the thermal use of the underground containing environmental aspects, basic requirements of components, and installation techniques. The VDI worked together with experts from neighboring countries like Switzerland, Austria, and Netherlands aiming to exchange experiences and to harmonize and facilitate international rule making. The VDI documents are also meant to be used as basic documents in the drawing up of European or international rules in harmonization with institutions responsible for these rules (Sanner, 2007). The EU wide review on policy shows that great differences between countries exist. Often these are due to differences in the physical environment and other uses of groundwater. But also the fact that SGE is a relatively new technology and there is not a clear EU platform yet to exchange experiences, plays a role. A global review on SGE policy was carried out by Hähnlein *et al.* (2010a) which showed that great differences exist worldwide and only few countries have specific legislation dealing with SGE.

2.4 DISCUSSION: SUBSURFACE TECHNOLOGY DEVELOPMENT AND REGULATION

From the presented overview on risks and policies relating to SGE we can identify two developments that require further attention. First, new technologies are developed by one sector without considering the impacts on other sectors. This not only poses potential threats to groundwater resources and the subsequent drinking water supply, but also generates conflicts with other potential users of the underground. This underscores the need for a cross-sectoral approach that balances all interests and for the development of master plans of all subsurface activities, not only drinking water and SGE. Traditionally, planning of underground works is done on a single-project basis with little consideration of other future potential uses. This single sector approach can produce interference between uses. Identifying the potential interferences and synergies between different subsurface activities (Table 2.2) early in the planning stages of regional (re-)developments, can avoid negative interferences between activities and allow for innovative opportunities for sustainable use of the subsurface, for example using waste heat from metro lines for heating buildings (Maire *et al.* 2006). Another interesting example is that of ATES and BTES systems being within each other's zones of influence. The notion that BTES systems could recover (steal) heat injected by an ATES system was one of the drivers to allow regulators to establish the earlier mentioned interference regions. A recent study by Drijver and Wennekes (2013) showed however that this effect was in most cases <10% and could be both positive and negative. Spatial subsurface planning is therefore required to minimize negative interference or in combining subsurface activities to achieve a greater mutual benefit.

Table 2.2 Matrix indicating interference between various subsurface functions.

	ATES	BTES	Deep Geothermal	Water supply	Infrastructure	Shallow geo-material (sand, gravel)	Deep geo-material (salts)	CO_2 storage	Gas-storage	Water storage
ATES	±	±	+	-	-	+	+	+	+	–
BTES		±	+	-	+	+	+	+	+	+
Deep Geothermal			–	+	+	+	–	–	–	+
Water supply				–	+	–	+	+	+	–
Infrastructure					±	±	+	+	+	-
Shallow geomaterials						–	+	+	+	-
Deep geomaterials							–	±	±	+
CO_2 storage								–	–	+
Gas storage									–	+
Water storage										–

– Subsurface functions will always conflict

\- Potentially conflicting subsurface functions

\+ Subsurface functions that almost never conflict with each other

± Potentially conflicting subsurface functions but mutual benefits can be achieved

Overall however, the speed of development and implementation of SGE systems has not been matched by the responsible policy makers. Regulators need to be aware of the emerging technologies and the activities of first movers on the market. New developments should be considered and weighed for their potential adverse or positive impacts on other stakeholder groups. Policy makers should strive to find a balance between setting unnecessary barriers to the development and implementation of new technology, and the protection of important issues such as safety as well as health and economic interests. The initial survey presented here on SGE implementation and regulation in the EU showed that many Member States are unaware of the activities within their territory or are unfamiliar with the concept at all. As a result, many countries do not have any regulation in place to deal with SGE activities. This resulted in a situation where regulation lags behind development and where unforeseen risks on the underground ecosystem and drinking water resources go unchecked.

2.5 CONCLUSIONS

Given the accelerating construction rate of SGE systems in the Netherlands and Europe, a growing number of systems is expected in the vicinity of drinking water well fields and other subsurface interests. This clearly indicates that water supply and energy supply are increasingly becoming each other's competitors for subsurface space. Although the risks that SGE can have on drinking water production are widely known, a comprehensive strategy to manage risks and monitor adverse impacts is currently lacking, other than a general restriction to apply SGE outside groundwater protection zones and other strategic groundwater reservoirs, and to adhere to a maximum injection temperature of 25°C.

The lack of a clear and scientifically underpinned risk oriented management strategy implies that potentially unwanted risks might currently be taken at vulnerable locations such as near well fields used for drinking water production, whilst at other sites the application of SGE is avoided without proper reasons. We have recognized three main issues that should be addressed to secure sustainable application of SGE: (1) Scientific research is required to further elucidate the impacts of SGE on groundwater; (2) Cross-sectoral subsurface planning is required to minimize negative conflicts between SGE and other subsurface interests; (3) EU wide guidelines and standards are required for quality assurance and control when installing a SGE system.

Chapter 3

A field and modeling study of the impacts of aquifer thermal energy storage on groundwater quality

ABSTRACT

In this chapter, chemical and microbiological water quality data are presented from an aquifer thermal energy storage (ATES) system located 570 m from a public water supply well field in the south of the Netherlands. The chemical data showed that the groundwater circulation by the ATES system impacted on chemical groundwater quality by introducing shallow groundwater with a different chemical composition at greater depth. This hypothesis was confirmed by numerical groundwater flow and solute transport modeling. The observed concentration changes were however sufficiently small to keep groundwater suitable for drinking water production. Microbiological data showed that groundwater circulated in the ATES system contained bacteria that with standard culture tests are characterized as fecal indicator bacteria. Molecular identification however showed that most of these organisms were not related to a fecal contamination and possibly naturally present in the aquifer. Two samples contained the fecal bacterium *Clostridium perfringens* but the presence was not linked to ATES operation and may be a relic of monitoring well construction. The presence of *Clostridium perfringens* did not form a hygienic risk because of the sufficient travel time between the ATES wells and the public supply well field.

3.1 INTRODUCTION

The exponential growth in the number of shallow geothermal energy (SGE) systems has led to an increasing number of sites where SGE systems are realized just outside the groundwater protection

*Based on:

*Bonte M., Stuyfzand P. J., van den Berg G. A. and Hijnen W. A. (2011) Effects of aquifer thermal energy storage on groundwater quality and the consequences for drinking water production: a case study from The Netherlands. Water Science Technology, **63**(9), 1922–1931.*

*Bonte M., van Breukelen B. M. and Stuyfzand P. J. (2013) Environmental impacts of aquifer thermal energy storage investigated by field and laboratory experiments. Journal of Water and Climate Change, **4**(2), 77–89.*

Hijnen W. A., Brouwer-Hanzens A., Bonte M. and Heijnen. L. (2011) Fecal contamination of groundwater at an ATES system in Eindhoven. KWR2011.107. 12 p.

zones of public supply well fields (Figure 1.3). This often leads to discussions between Dutch drinking water companies who want to protect their groundwater resources (contributing around 60% of the national drinking water supply), and project developers who want to use groundwater for SGE. This discussion is clouded by many uncertainties on the effects of SGE systems on drinking water production in general and groundwater quality in particular. This uncertainty is partially due to the lack of research using field data on the effects of low temperature SGE systems on groundwater quality under Dutch aquifer conditions.

In Chapter 2, we described that most published research focused on operational aspects, such as scaling due to mineral precipitation occurring at high temperature systems, and was based on either laboratory experiments (Griffioen & Appelo, 1993), chemical equilibrium modeling (Palmer & Cherry, 1984) or a combination of both (Arning *et al.* 2006a). Very few field data of SGE systems have been reported in peer reviewed literature to date. Monitoring data for water quality has, to our knowledge, not been reported for BTES systems, and also data on typical low temperature ATES systems as operated in the Netherlands have not been reported in peer reviewed literature previously and only sporadically in grey literature (Dinkla *et al.* 2012).

The main questions in this chapter are (i) whether ATES and drinking water production can operate in the same aquifer at limited distance without negatively influencing each other, (ii) what the effects are of ATES operation on chemical and microbiological groundwater quality and (iii) which processes dominate changes in groundwater quality. We used water quality and temperature monitoring data from an active ATES system and nearby groundwater monitoring wells to investigate the impacts on chemical and microbiological water quality. A selection of observation data were analyzed quantitatively using a groundwater flow and solute transport model.

3.2 SITE DESCRIPTION

The ATES system under investigation was constructed in 2003 and started operation in January 2005. The ATES system is located in the south of the city of Eindhoven, between highway A2 and the Dommel River (refer Figure 3.1). The system is located between two Public Supply Well Fields (PSWF), both operated by Brabant Water: Aalsterweg about 650 m to the east and Klotputten approximately 570 m to the west. The ATES system is partly located within the groundwater protection zone of the first mentioned PSWF. The groundwater protection zone is defined by the area where infiltrating rainfall will reach the extraction wells within a period of 25 years. In this area, activities that may compromise water quality are prohibited. In most cases, this includes the establishment of an ATES system. In this case however, authorities allowed the system under the condition that it would serve as a research site to investigate the effects of ATES on groundwater quality. This location was selected, and not a site further away from the existing PSWF, by virtue of the presence of groundwater monitoring wells with a background groundwater quality data set. Also shown in Figure 3.1, is the 60 day protection zone, which is the area in the aquifer where groundwater will reach the extraction wells within 60 days.

A hydrogeological cross section showing both ATES wells and the PSWF is shown in Figure 3.2. The ATES system is located in sections of the Sterksel aquifer consisting of coarse sand between 5 and 55 m-MSL (meters below mean sea level). The systems consists of 10 ‘cold’ wells (injecting cold water) and 10 ‘warm’ wells concentrated in four well clusters. During an average climatic year, the design groundwater circulation of 2.6×10^6 m^3 generates 23,100 MWh. The design average injection temperatures are 7 and 16°C in the cold and warm wells, respectively. Temperature logs taken at monitoring wells uninfluenced by the ATES system show the natural groundwater temperature in the Sterksel aquifer ranges between 10.5 and 12°C.

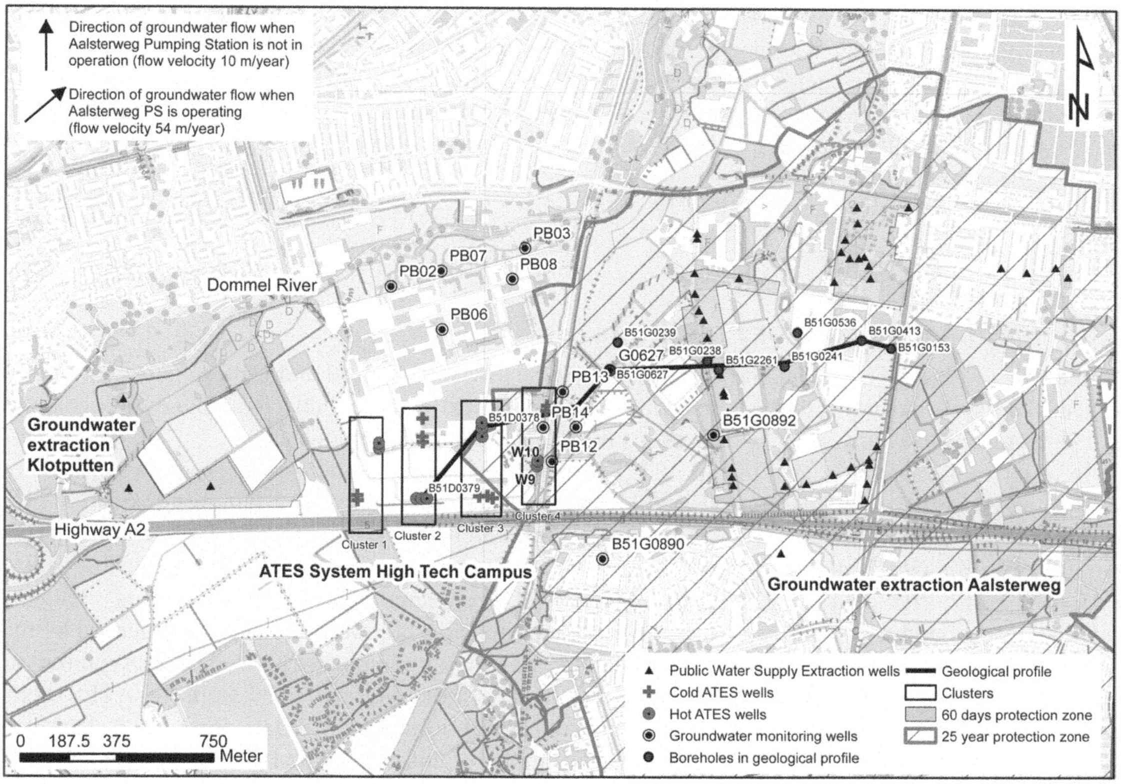

Figure 3.1 Location of the ATES system, monitoring wells and the two public supply well fields of Brabant Water.

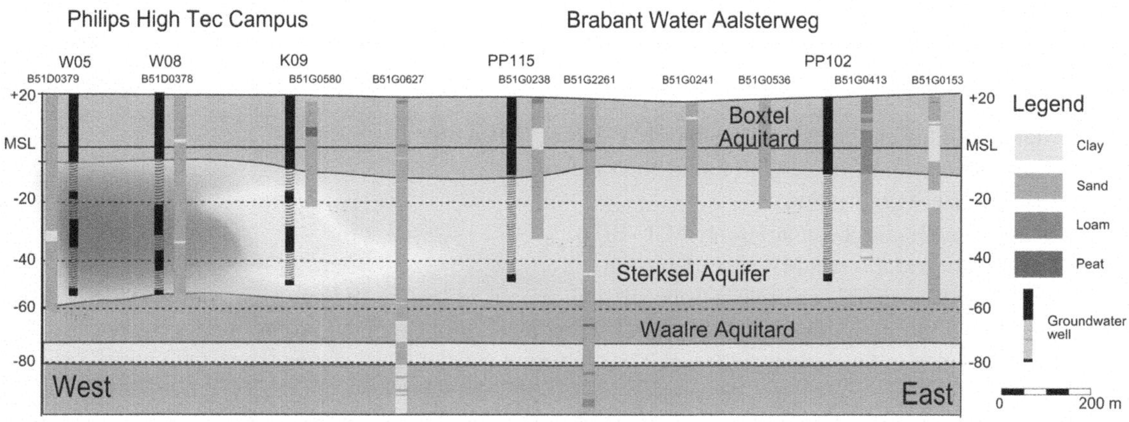

Figure 3.2 Hydrogeological cross section over the ATES system and the Aalsterweg public water supply well field. Position of cross section is shown in Figure 3.1.

The Aalsterweg PSWF extracts groundwater from both the Sterksel aquifer (licence of 5×10^6 m^3/year) and the much deeper Kiezeloolite aquifer. The Klotputten PSWF extracts groundwater from the much deeper Kiezeloolite aquifer (190 to 340 m-MSL). Groundwater head monitoring data were used to derive the direction of groundwater flow through triangulation. This showed flow velocities and direction of 54 m/year towards the north east when Aalsterweg is in operation and 10 m/year to the north in case no pumping occurs at Aalsterweg.

3.3 METHODS

3.3.1 Field and laboratory methods

Monitoring of the ATES system was carried out by a Supervisory Control and Data Acquisition (Scada) system. This system controls the acclimatization and logs the extracted and injected groundwater volumes and temperature on a daily basis and calculates the corresponding delivered energy.

In order to assess the effects of the ATES system, we compared changes in groundwater quality in two ATES wells, with the natural spatial and temporal variability in groundwater quality in a number of reference wells. For this purpose, groundwater was sampled from two 'warm' ATES production wells (W09 and W10) and several groundwater monitoring wells (PB2, PB3, PB5, PB6, PB7, PB8, PB12, PB13 and G0627), shown in Figure 3.1. From temperature readings and mass balance calculations, groundwater samples drawn from the monitoring wells were shown not influenced by the ATES. The chemical composition of these samples was therefore considered to be representative of the ambient groundwater quality. Microbiological samples were only collected from the ATES wells and not from the groundwater monitoring wells. We used microbiological data from the Aalsterweg PSWF as a reference for the microbiological water quality of ambient groundwater for the monitoring period 2005–2009.

Samples were analyzed for major ions, a selection of trace metals, dissolved organic carbon (DOC), nitrate, ammonium and a selection of microbiological parameters. The location of the monitoring wells is presented in Figure 3.1. The water of the two warm wells of cluster 4 was sampled from a piezometer located in the gravel pack screened between respectively 35.8–37.9 m-MSL (warm well W09) and 50.0–52.0 m-MSL (warm well W10). The wells were sampled with a rather irregular frequency due to the irregular operation of the ATES system. The average frequency was bimonthly over the period 2005 to 2009. All sampling and laboratory analyses were done by Waterlaboratorium Zuid (Breda, the Netherlands). Laboratory analyses included:

- Major cat-ions and metals with either ICP-MS or ICP-AES;
- Major anions and silicate with flame photometry;
- Bicarbonate with titration;
- Methane with gas chromatography (GC-FID);
- DOC with oxidation and IR light detection;
- *Escherichia coli* (*E. coli*), Enterococci and spores of sulfite-reducing clostridia (SSRC) with membrane filtration (methods ISO 9308-1, ISO 7899-2 and Dutch NEN6567, respectively);
- Heterotrophic colony count at 22 and 37°C (Method ISO 6222).

Following the sampling campaign between 2005–2009, an additional sampling round was carried out in 2011 on both ATES wells and the monitoring wells sampled in earlier rounds, to verify the presence of fecal bacteria and identify them. In addition two wells located further away from the ATES system were sampled, serving as additional background monitoring wells (B51G0890 and B51G0892 in Figure 3.1), and one monitoring well realized in 2011 within the zone of influence of the ATES system (PB14). Sampling was done by collecting 40 L groundwater with an optimized sampling protocol to avoid microbial contamination. This large volume sample was split into 5 L samples for the microbiological analysis on *E. coli*, Enterococci and SSRC. The influence of iron precipitation during filtration of these samples on the recovery was assessed by standard addition of *E. coli* in a separate sample. The recovery of the E. coli enumeration was 26%. The results of microbial analysis were corrected for this (low) recovery percentage.

Presumptive positive colonies were sampled for DNA followed by sequencing of PCR-amplified bacterial 16S rRNA genes (approximately 1250 nucleotides) for Enterococci and SSRC. The assembled sequences were compared for further identification with sequences deposited in GenBank using BLAST (Altschul *et al.* 1990).

3.3.2 Numerical modeling of the field data

The groundwater flow conditions and resulting sulfate concentrations observed in a selection of the ATES wells were simulated using Modflow (Harbaugh *et al.* 2000) for groundwater flow coupled with MT3DMS (Zheng, 1990) for solute transport. Sulfate was chosen here as a key parameter showing the largest deviations from observed ambient concentrations (further detailed in results and discussion section). This modeling framework considers solute transport by advection (based on the flow field calculated by Modflow) and dispersion. The coupling between the extraction and injection well and mixing is done by assigning a negative input solute concentration (CSS) to the MT3DMS sinks and source packages, according to (Zheng, 2010):

$$\mathrm{CSS} = -\mathrm{NCOL} \times \mathrm{NROW} \times (\mathrm{K} - 1) + \mathrm{NCOL} \times (\mathrm{I} - 1) + \mathrm{J} \qquad (3.1)$$

where NCOL and NROW are the total numbers of columns and rows in the model, respectively, and K, I, J are the layer, row, and column indexes of the model cell in which the extraction cell is located whose concentration is to be used as the input concentration of the recirculation well. The method allows for using 1 grid cell as the extraction well coupled with the injection well. In the model the extraction well is set in all layers in the Sterksel aquifer by assigning the extraction cell in the top cell in the Sterksel aquifer and giving underlying cells in the aquifer a very high conductivity (10^6 m/d). The observed SO_4 variation with depth is used to describe the initial conditions for each layer. Reactive transport was not included in the model (although this is possible in MT3DMS), because the processes included already provided sufficient detail to describe the main trends in observed sulfate concentrations. This does not mean that reactive processes can be completely ruled out, merely that the observation data and temperature differences at the ATES site were insufficient to quantify the temperature dependence of this key biochemical process. In Chapter 5, we used anaerobic laboratory experiments to investigate this process in more detail.

The model's hydrogeological setup is based on the geological cross section presented in Figure 3.2. The Aalsterweg public supply well field comprises of two north-south trending well fields of roughly 1 km length located 600 m east from the ATES system. The effect of this well field on the ATES system is simulated by a constant head boundary at the east and west boundaries of model domain which sets the hydrological gradient imposed by the well field. Table 3.1 provides further details on the model discretization, boundary and initial conditions.

Table 3.1 Summary of discretization, model parameters, and boundary and initial conditions used for Modflow/MT3DMS modeling.

Model aspect	Model value
Model discretization	
Model domain	1×1 km^2
Horizontal discretization	40×40 m^2 at the boundaries 5×5 m^2 at the center of the model near the ATES system
Vertical discretization	16 layers, representing Boxtel semi-confining layer (4 layers) and Sterksel aquifer (12 layers)
Flow parameters (Modflow)	
Horizontal conductivity (K_h)	Upper Boxtel (+20 to 0 m-mean sea level, MSL): 20 m/d (medium fine sand)
Based on values of the Dutch geological database (REGIS) accessible via www.dinoloket.nl	Lower Boxtel (0 to –5 m MSL): 0.05 m/day (sandy clay) Sterksel (–5 to –55 m MSL): 30 m/d (medium coarse sand)
Vertical conductivity (K_v) Also from REGIS	Upper Boxtel: 10 m/d Lower Boxtel: 0.01 m/day Sterksel: 15 m/d
Porosity (n)	0.35
Specific storativity (S_s)	10^{-4} m^{-1}
Boundary conditions	Ambient flow field with hydraulic gradient towards the Brabant Water Aalsterweg simulated by constant head boundaries at east and west boundaries of the model simulating an ambient flow velocity of 55 m/year.
ATES system	Simulated with injection and extraction wells (based on average monthly extraction rates of 25 m^3/hour)
Initial conditions	Bases on steady state results
Time discretization	Total simulation time: 3 years. Each year comprises: – 100 days extraction from warm well, injection in cold well during winter – 80 days storage period – 100 days extraction from cold well, injection in warm well during winter – 80 days storage period
Transport parameters (MT3DMS)	
Longitudinal dispersivity	2.5 m
Horizontal and transverse dispersivity	0.25 m
Diffusion	ignored
Advection	Third-order TVD (total-variation-diminishing) scheme (time discretization based on a Courant number of 0.75)
Boundary conditions	Solute sinks at extraction wells Solute sources at injection wells, based on average extraction sulfate concentration calculated by MT3DMS by use of Equation (3.1). Fixed concentration at all outer model boundaries (based on observed SO_4 gradient): – Boxtel aquifer: 40 mg SO_4 – Upper part of Sterksel aquifer (5 to 25 m BSL): a linear gradient from 40 to 5 mg SO_4 – Lower part of Sterksel aquifer (25 to 55 m BSL): 5 mg SO_4
Initial concentration	Same depth distribution as boundary conditions

3.4 RESULTS

3.4.1 Flow and temperature data

Temperature monitoring data showed that in summer, water temperature in the warm wells increased up to 28°C; while in winter the temperature in the cold wells decreased to a minimum value of 6°C (Figure 3.3). During July 2005 and August 2006 the cold and warm wells of clusters 1 and 2 reported similar temperatures in excess of 20°C. This is unlikely to be representative of the temperature of the groundwater. The temperature sensor may have been influenced by atmospheric heat or heat generated by the pump. Ignoring the periods that the temperature sensor may have been influenced by external factors, we see that the temperature difference between the warm and cold wells ranges between 3 to 5°C, which is less than the design temperature difference of 9°C. It reaches a temperature of 7°C during injection in winter and of 16–18°C during injection in summer.

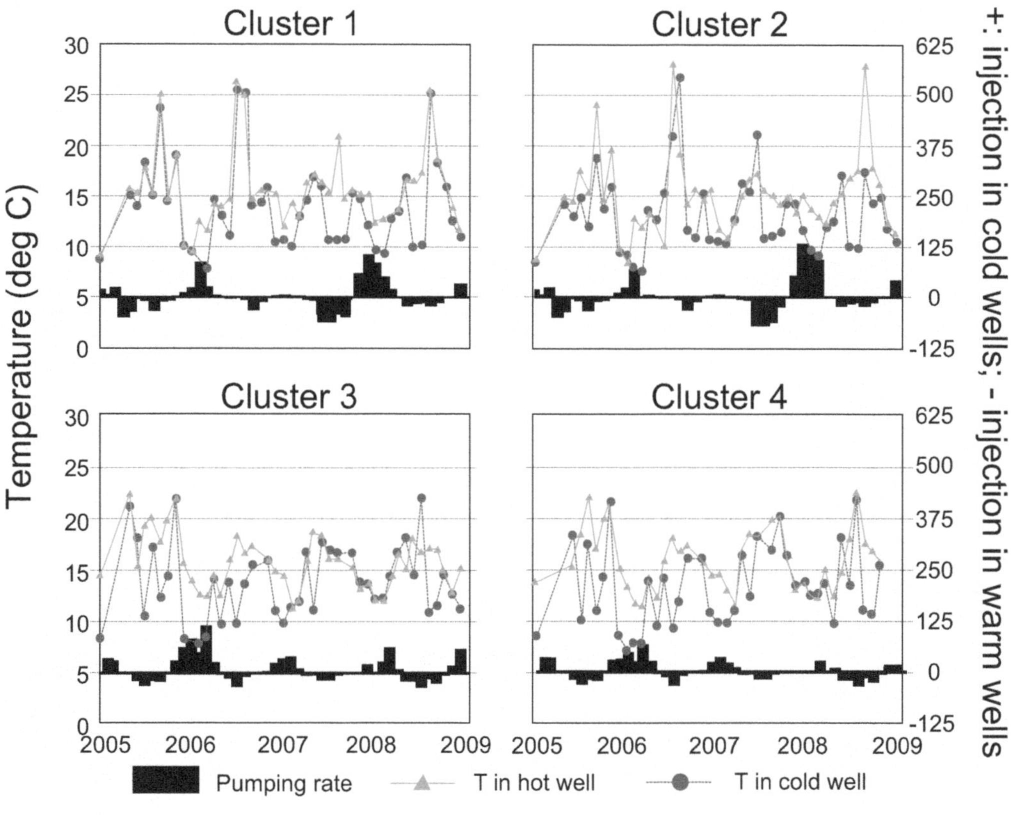

Figure 3.3 Pumping rates and water temperatures in ATES wells from the four well clusters.

3.4.2 Ambient chemical groundwater quality

The ambient chemical depth profiles for NO_3, dissolved Fe, SO_4, and CH_4 shown in Figure 3.4 reveal a typical redox zonation as described by Appelo and Postma (2005) with a transition from post-oxic groundwater to a depth of around 20 m-MSL to methanogenic groundwater below 30 m-MSL. Possible sources for the elevated concentrations in shallow groundwater include historic changes in atmospheric deposition (for SO_4 and NH_4), road de-icing salt (Na & Cl), leaky sewers (all), and fertilization (NO_3) (Broers & van der Grift, 2004; Appelo & Postma, 2005).

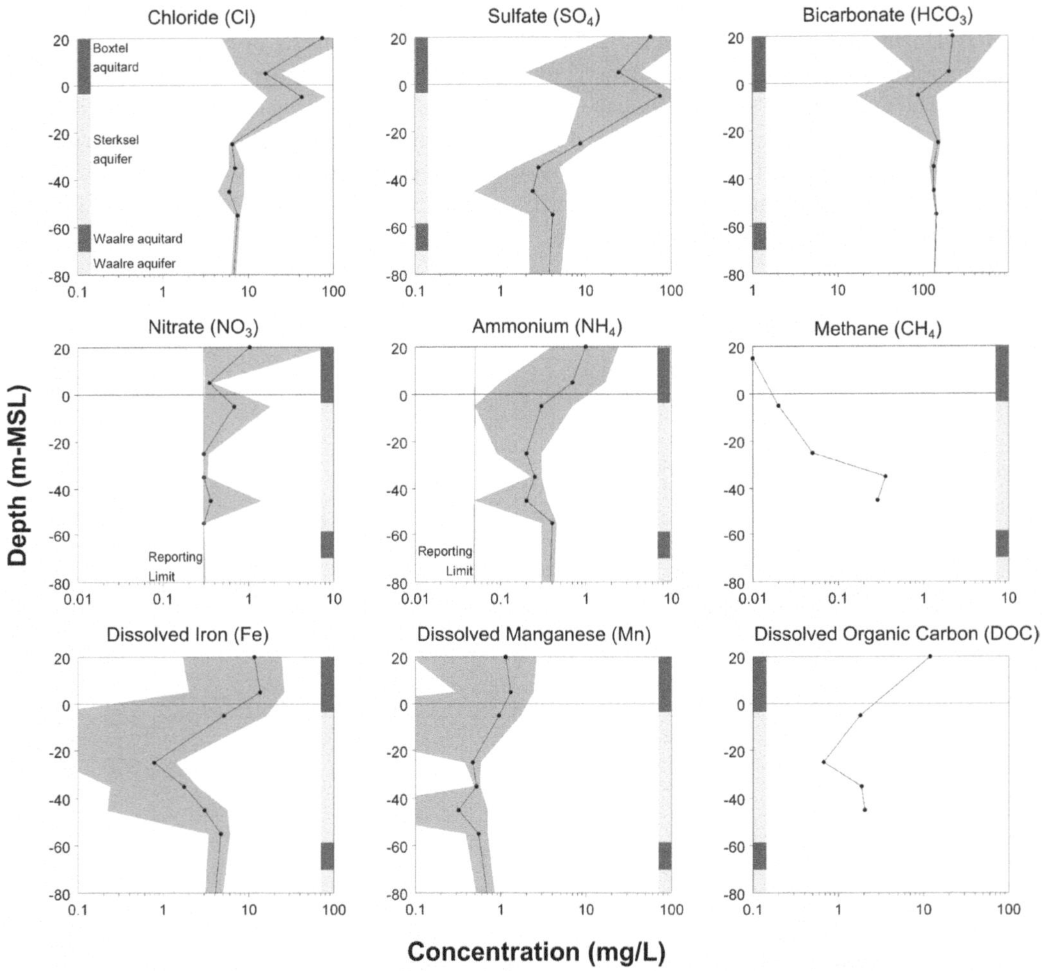

Figure 3.4 Hydrochemical depth profiles showing minimum and maximum concentration in shaded area surrounding the average concentration indicated as a black line. Profiles are based on data from monitoring wells PB2, PB3, PB5, PB6, PB7, PB8, PB12, PB13 and G0627.

Nitrate present in shallow groundwater is likely to be reduced by reactive geochemical sediments. Geochemical data of aquifer material from the Sterksel aquifer at the Langerak pumping station indicate that pyrite and organic matter and to a lesser extent siderite are the driving reductants for this process (Stuyfzand, 1998; Stuyfzand, 2001; Hartog *et al.* 2002; Appelo & Postma, 2005). This shows that while the shallow groundwater in the aquifer is likely to be influenced by changes in land use and (urban) pollution at the land surface, the deeper section of the aquifer is better protected from polluting activities at land surface but highly reduced and containing methane.

3.4.3 ATES chemical water quality

A selection of time series for the ATES water quality is plotted in Figure 3.5. Each graph includes a bar chart showing the extraction rate of the ATES system and two lines showing the minimum and maximum concentration for a given element in the ambient groundwater at a given depth, as derived from the data shown in Figure 3.4. For most parameters, the time series of the two sampling points show nearly identical data during periods of injection but they differ from each other during periods of extraction. This is explained by a common source of water during injection (i.e., the cold wells) and the difference in origin or extraction depth during periods of extraction.

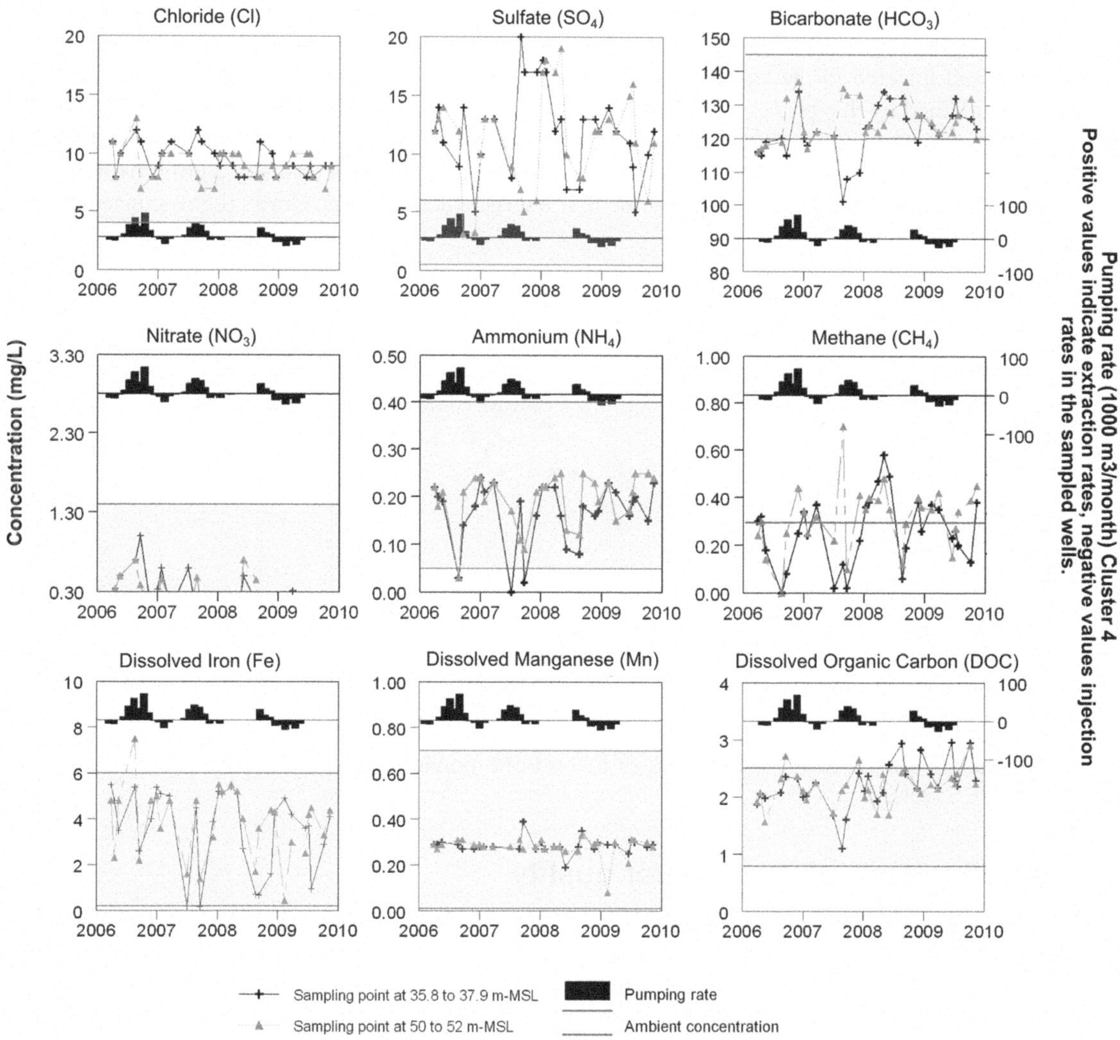

Figure 3.5 Concentration time series of ATES water from W9 (35.8 to 37.9 m-MSL) and W10 (50 to 52 m-MSL), as sampled from a piezometer in the gravel pack of the warm well. Ambient range in concentration is shown with shading as derived from Figure 3.4 Methane was analyzed once.

The variation in the conservative element chloride is in the order of only a few mg L^{-1}. However, the concentration is higher than observed in ambient groundwater. The SO_4 maximum concentration of the ATES water is 20 mg L^{-1}, which is about four times higher than the maximum observed in ambient groundwater at the same depth. The SO_4 concentration is still well below the maximum admissible concentration (MAC) for drinking water of 150 mg L^{-1}. The nitrate concentrations in the ATES water (0.3 to 1.0) are just above the limit of quantification (LOQ) of 0.3 mg NO_3 L^{-1}. These values are in the same order of magnitude as observed in the ambient groundwater at this depth (0.3 to 1.4 mg NO_3 L^{-1}). The dissolved organic carbon (DOC) data show a slight increase over time: from around 2 mg L^{-1} between 2005 to 2007 to maximum values of 2.5 to 3 mg L^{-1} between 2007 and 2009. The iron data show a cyclic pattern with high iron concentrations during injection and low concentrations during storage and extraction.

The most logical source for the elevated Cl concentrations in the ATES water is a fraction of shallow groundwater which is extracted in the upper part of the filter of the cold well, mixed with deeper groundwater, and injected in the warm well. In order to test whether the varying sulfate concentrations can also be explained by this process SO_4 was plotted against Cl (Figure 3.6) which shows that the ATES samples plot on a mixing line between the samples collected in the upper 20 m of the Sterksel aquifer and the lower part of the Sterksel Aquifer. As there are no geochemical Cl sources in the aquifer (which holds for most Dutch shallow aquifers) we can assume that a large fraction of the elevated SO_4 concentrations can be explained by conservative mixing.

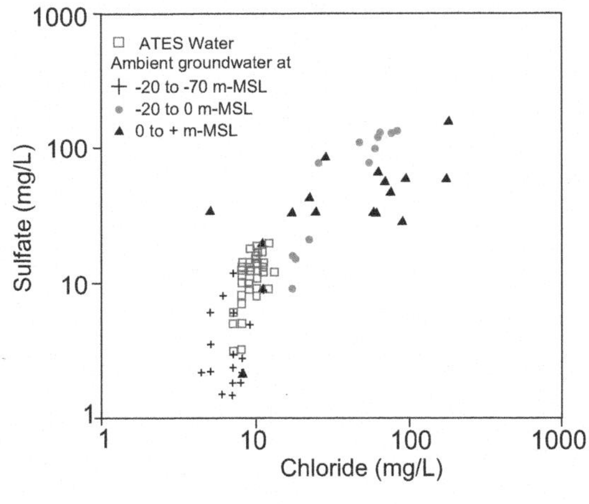

Figure 3.6 Sulfate versus chloride for samples taken from monitoring wells (ambient groundwater) and samples of ATES water.

3.4.4 ATES microbiological water quality

Table 3.2 presents a summary of the results of the microbiological analyses of water sampled from the ATES wells and the ambient concentrations as analyzed from groundwater sampled from the wells at the Aalsterweg PSWF. The drinking water directive (98/83/EC) determines that water used for human consumption must be free of *E. coli* and Enterococci (0 CFU/100 mL). The data show that fecal indicator bacteria *E. coli*, Enterococci, and SSRC were detected in the ATES samples while none of ambient groundwater samples revealed these bacteria. We note that while most positive analyses of *E-coli and* Enterococci were in the range of 1 to 5 CFU/100 mL, one sample reported counts of *E-coli* and Enterococci of 120 and 9 CFU/100 mL, respectively (individual data not shown in Table 3.2). The colony count data showed a higher microbiological activity in the underground near the ATES wells compared to the activity under ambient groundwater conditions, especially for bacteria which grow at higher temperatures.

Table 3.2 Summary of the microbiological data for ambient groundwater and ATES water collected between 2005 and 2009.

Parameter	Ambient groundwater Aalsterweg PSWF		ATES well W9		ATES well W10	
	Pos	*n*	Pos	*n*	Pos	*n*
E. coli per 100 mL	0	118	2	29	1	29
Enterococci per 100 mL	0	13	6	29	4	29
SSRC per 100 mL	0	10	19	29	17	29
	μ	max	μ	max	μ	max
Colony count 22°C (CFU/mL)	195	8.1×10^3	1.6×10^3	1.1×10^4	2.4×10^3	8.4×10^3
Colony count 37°C (CFU/mL)	1.3	12	454	4.0×10^3	435	3×10^3

Note 1: Abbreviations used: Pos, Number of samples analyzed positive for the occurrence of bacteria; *n*, total number of samples analyzed; μ, average colony count; max, highest colony count; SSRC, spores of sulfite reducing clostridia; CFU, colony forming unit.

In verification monitoring round in 2011, in one 5 L water sample from the ATES system coliforms were detected. These colonies however tested negatively for *E. coli*. Thus, the observations of *E-coli* (3 observations) in the 2005–2009 monitoring series, were not confirmed in the 2011 higher volume sampling round. Contrary to the absence of *E-coli*, the results of the first monitoring campaign for enterococci and SSRC were confirmed. In all groundwater samples, SSRC and enterococci were found at concentrations of 0.01–2.2 CFU/100 mL. These observations are in agreement with the result of the first set of data where enterococci and SSRC were regularly observed in the 100 mL samples (Table 3.2).

Molecular identification however showed that DNA sequences derived from the enterococci colonies have the greatest resemblance (99–100%) with *Staphylococcus pasteuri* or *Staphylococcus warneri*, which are not of fecal origin. DNA sequences derived from the SSRC medium showed the greatest resemblance (95–100%) with *Clostridium frigidicarni, Paenibacillus konsidensis, Bacillus circulans/ Paenibacillus spp, Paenibacillus terrae/Paenibacillus spp, Clostridium subterminale/C. thiosulfatireducens, Pseudomonas putida, Clostridium ruminantium, Clostridium bifermentans/C. difficile,* and *Paenibacillus caespitis/ Paenibacillus spp.* These are frequently observed in soil environments and therefore assumed to be not an indication of a significant fecal contamination. In two water samples collected from monitoring wells PB14 (north of W10, Figure 3.1) and B51G0890 (reference) sequences were identified closest matching *Clostridium perfringens* (99–100%) which is a bacterium of fecal origin. The presence of the relatively persistent spores of *C. perfringens* indicates a fecal contamination source that is not active any more.

3.4.5 Numerical modeling of field data

A more quantitative analysis of the mixing process inferred to be responsible for the cyclic behavior is provided by groundwater flow and solute transport modeling. Cross sections of the modeling results (Figure 3.7) clearly show the extraction, mixing, and injection during ATES operation which disturbs the natural stratified sulfate distribution with high sulfate concentrations in the Boxtel aquifer and upper Sterksel aquifer, and low concentrations in the lower Sterksel aquifer, generating plumes of 'injected ATES water' with an average concentration of both aquifer layers. The close vicinity of the PSWF causes a relatively high groundwater flow velocity. This causes the injected plumes of ATES water to drift away with the ambient groundwater flow towards the PSWF while somewhat fading out due to dispersion. Also visible in the cross sections, is that the drifting plume of injected ATES water, actually flows from one ATES well to the adjacent coupled ATES well, thus causing a second jump in the breakthrough of mixed ATES water. This is highlighted with circles in Figure 3.8, displaying this second breakthrough in the observed data.

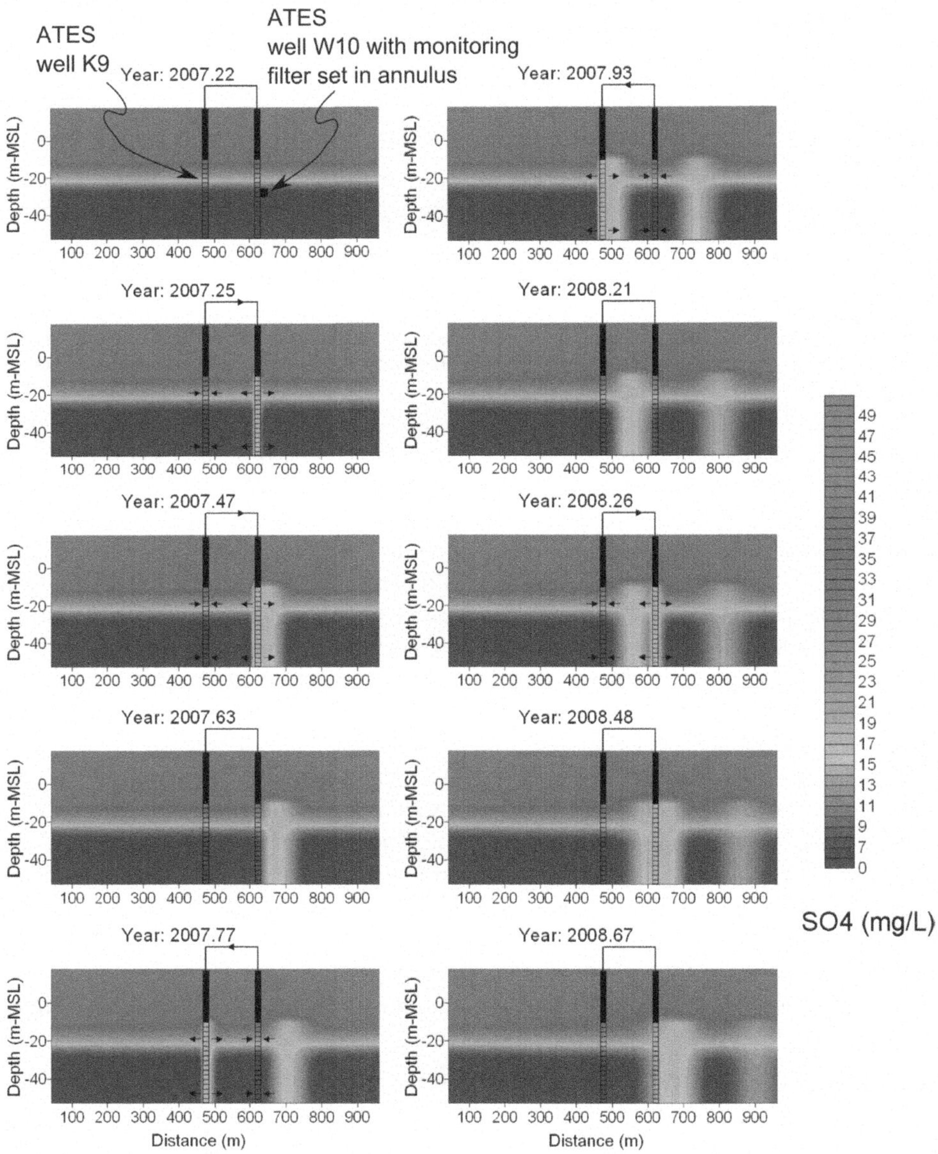

Figure 3.7 Simulated sulfate concentrations (shading) during several periods over the 3 year simulation period. Two ATES wells (K9 and W10) are shown, top left panel shows location of the monitoring screen set in the annulus of W10. The observed and simulated SO_4 concentrations for this monitoring screen are shown in Figure 3.8.

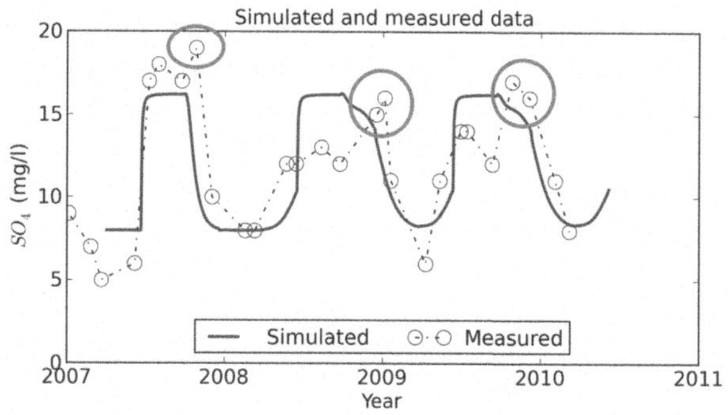

Figure 3.8 Measured and observed sulfate concentrations in ATES well W10 with screen between −50 and −52 m-MSL over the period 2007–2010. The circles show the second breakthrough of ATES water from the adjacent ATES well (see to text for further description).

The agreement between observed and simulated sulfate concentrations (Figure 3.8) indicates that the process of extraction, mixing, and injection can adequately describe the observed oscillations in sulfate concentrations. The observed and modeled sulfate concentrations show that the injected, heated, and chilled water actually floats away from the ATES wells before being recovered in the subsequent season. Temperature data in the ATES warm and cold wells presented in Figure 3.3 show that the temperature in the 'warm' well is increased during summer and decreased during injection in the 'cold' well during winter. When extraction starts in the subsequent season, the temperature in the ATES well has returned to near ambient conditions confirming that the ATES water has drifted away before it can be recovered. This is a-typical for most areas in the Netherlands where groundwater flow velocities are quite low and indicates that areas near existing well fields are not very suitable for ATES systems because stored heat is likely to drift away before it can be recovered.

3.5 DISCUSSION

Overall, the data and modeling of the field site showed that ATES operation changed the natural water quality zoning in the aquifer. This water quality zonation is likely the result of the typical sequence of redox processes occurring in the aquifer (denitrification, iron-reduction, sulfate-reduction and methanogenesis) (Appelo & Postma, 2005) and changes in historic input of solutes with recharge (Broers & van der Grift, 2004). The re-distribution of electron acceptors may impact on the local redox conditions and the microbial communities present. Redox conditions are important for a number of geochemical processes, for example the sorption and mobility of trace metals (Lovley & Coates, 2000; Appelo & Postma, 2005; Wallis *et al.* 2011) or the degradation of organic micro-pollutants (Greskowiak *et al.* 2006). The redistribution of electron acceptors may cause shifting redox conditions and associated water quality changes. These secondary redox impacts were however not observed in this ATES system as the injected water drifted away before it was recovered again.

In a more generic sense, however, it can be stated that the vulnerability of the public supply well field is increased because shallow water, more influenced by human activity is brought to a depth where it can travel through coarser sediment layers and reach the public supply well field more rapidly. In addition to this, the capacity of the aquifer to buffer contaminants (sorption of organic contaminants to organic matter or metals to iron-oxides, or by reducing introduced oxidants) may decrease with depth as the aquifer shows a downward coarsening of sand texture.

The absence of pathogens is of paramount importance to the suitability of groundwater for drinking water supply, especially in the Netherlands where chlorination or other disinfection treatment methods are hardly used in preparing drinking water from groundwater (Smeets *et al.* 2009). The elevated counts of the indicator bacteria (*E. coli*) point to a recent fecal contamination which was however not confirmed in a monitoring campaign with optimized sampling of large volumes indicating that either system was only temporarily open to fecal contamination or that the sample was contaminated. Molecular identification showed that most SSRC and all enterococci are not of fecal origin but part of the naturally occurring bacteria population, and therefore designated as false positives. This highlights the need for colony identification of presumptive fecal colonies observed in the culture methods commonly used in microbiological water control, especially when fecal indicator parameters are used in groundwater environments which provide a complex habitats for microbial communities (Griebler & Lueders, 2009). The natural occurrence of enterococci in soils was also noted in a recent extended review on enterococci as a fecal indicator bacterium (Byappanahalli *et al.* 2012).

The positive identifications of *Clostridium perfringens* and absence of other fecal bacteria, indicated that fecal contamination occurred in the past (Bisson & Cabelli, 1980). Spores of clostridia are very persistent and can survive very long times in natural environments (Medema *et al.*1997; Hijnen *et al.* 2009) A more recent study further shows that this indicator is only valid to point at contamination of excreta from non-herbivorous wildlife, human sewage, or excreta of dogs and cats in urban areas (Vierheilig *et al.* 2013). Because *Clostridium perfringens* is observed in both a monitoring well near an ATES well (PB14) and a reference monitoring well located far from the ATES system (B15G0890), the contamination is not likely to be related to the elevated temperatures caused by the ATES system. The fecal contamination of these monitoring wells may originate from the construction phase when the work was not carried out under hygienically safe conditions. This is however not specific for ATES and can occur in any monitoring well, groundwater extraction wells or BTES borehole, if drilled with unhygienic work water or drilling tools (Driscoll, 1986) or if the boreholes were not properly sealed to prevent inflow of surficial water (Knappett *et al.* 2012). Data from field and column experiments reported in literature show however that under Dutch aquifer conditions, a subsurface residence time ranging between 60 to 110 days is generally sufficient to reduce the presence of micro-organisms to an acceptable level (Hijnen *et al.* 2005, Wielen *et al.* 2008). Considering that the ATES system, and the monitoring wells where *Clostridium perfringens* were observed, are situated on the edge of the 25 year groundwater protection area, it is unlikely that the PSWF was at risk from the past microbiological contamination indicated by *Clostridium perfringens*.

Overall, for the current ATES system under investigation, solute concentrations remained below the MAC limits for drinking water. The fecal indicators that were identified do not form a likely threat as there is a sufficient travel time between the locations where these were observed and the extraction wells. We can conclude that there are no indications that groundwater became less suitable for drinking water production at this location. However, the data do indicate that ATES impacts on groundwater quality. The magnitude of the effects and processes described here on a PSWF's raw water quality depends on the ATES system design (well depths, injection temperature and production rate), aquifer conditions (groundwater quality gradients, conductivity and anisotropy), the PSWF setup (screen depth), land use and pollutant input. An unfavorable combination of system design and aquifer conditions may cause the impacts observed here to be more pronounced, making groundwater less suitable for drinking water production.

3.6 CONCLUSIONS

The chemical data showed that groundwater circulation by the ATES system impacted on chemical groundwater quality by introducing shallow groundwater with a different chemical composition at greater depth. This hypothesis was confirmed by numerical groundwater flow and solute transport modeling. The observed concentration changes were sufficiently small to keep groundwater suitable for drinking water production. Mixing by the ATES system may however change the vertical distribution of redox conditions which can, depending on site specific conditions, impact on the mobility of trace elements and degradation of organic contaminants or mobilize shallow anthropogenically influenced groundwater. This is added to the effect of temperature changes by storage of thermal energy. The combination of mixing and bubble drift however prevented an adequate analysis of the temperature induced changes and necessitated us to resort to laboratory experiments to focus on this aspect, described in the following chapters.

Microbiological data showed that groundwater circulated in the ATES system contained bacteria that with standard culture tests are characterized as fecal indicator bacteria. Molecular identification however showed that most of these organisms were not related to a fecal contamination and possibly naturally present in the aquifer. Two samples contained the fecal bacteria *C. perfringens* but the presence was not linked to ATES operation and may be a relic of the construction of the monitoring well. The presence of *C. perfringens* however formed no hygienic risk because of the sufficient travel time between the ATES wells and the public supply well field.

Chapter 4

Temperature-induced impacts on mobility of arsenic and other trace elements

ABSTRACT

In this chapter, we report the results of column experiments to assess the impacts of temperature variations (5°C, 11°C, 25°C and 60°C) on groundwater quality in anoxic reactive unconsolidated sandy sediments derived from an aquifer system widely used for drinking water production in the Netherlands. Our results showed that at 5°C no effects on water quality were observed compared to the reference of 11°C (in situ temperature). At 25°C, As concentrations were significantly increased and at 60°C, significant increases were observed in the pH and dissolved organic carbon (DOC), P, K, Si, As, Mo, V, B, and F concentrations. These elements should therefore be considered for water quality monitoring programs of shallow geothermal energy projects. No consistent temperature effects were observed for the trace elements Ba, Co, Cu, Ni, Pb, Zn, Eu, Ho, Sb, Sc, Yb, Ga, La, and Th, all of which were present in the sediment. The temperature-induced chemical effects were probably caused by (incongruent) dissolution of silicate minerals (K and Si), desorption from, and potentially reductive dissolution of, iron oxides (As, B, Mo, V, and possibly P, and DOC), and mineralization of sedimentary organic matter (DOC and P).

4.1 INTRODUCTION

In the previous chapter we presented the monitoring results from an ATES system located near a drinking water pumping station. The combination of mixing and bubble drift at the investigated site prevented us to carry out an adequate analysis of the temperature induced changes and necessitated to resort to laboratory experiments to focus on the thermal impacts on water quality. The behavior of trace elements and heavy metals was however not investigated in any of these aforementioned studies. In order to investigate the temperature driven hydrochemical changes, in particular the mobility of trace elements and heavy metals, we carried out anoxic column experiments at 5, 11, 25, and 60°C and analyzed a broad suite of chemical species, including major ions, trace elements, nutrients, and dissolved organic carbon. As stated in the review presented in Chapter 2, in particular the mobility of trace elements and heavy metals under changing temperatures in SGE systems is unknown and this was a particular focus of these experiments.

*Slightly modified from: Bonte M., van Breukelen B. M. and Stuyfzand P. J. (2013) Temperature-induced impacts on groundwater quality and arsenic mobility in anoxic aquifer sediments used for both drinking water and shallow geothermal energy production. Water Research, **47**(14), 5088–5100.*

The objectives of the research presented in this chapter were to (i) present an overview of the naturally occurring elements including trace metals that can be expected to be mobilized due to a temperature increase caused by ATES operation, (ii) identify the associated processes causing their mobilization, and (iii) provide guidelines for monitoring near shallow geothermal systems.

4.2 MATERIALS AND METHODS

4.2.1 Sediment collection, characterization, and geochemical analyses

Sediments and water used in the column experiments were collected in the Netherlands at sites in Helvoirt (sediment samples H1 and H2) and Scherpenzeel (sediment sample S) and consisted of unconsolidated anoxic sand from Sterksel formation. This formation is widely used for both drinking water production and ATES (Figure 4.1).

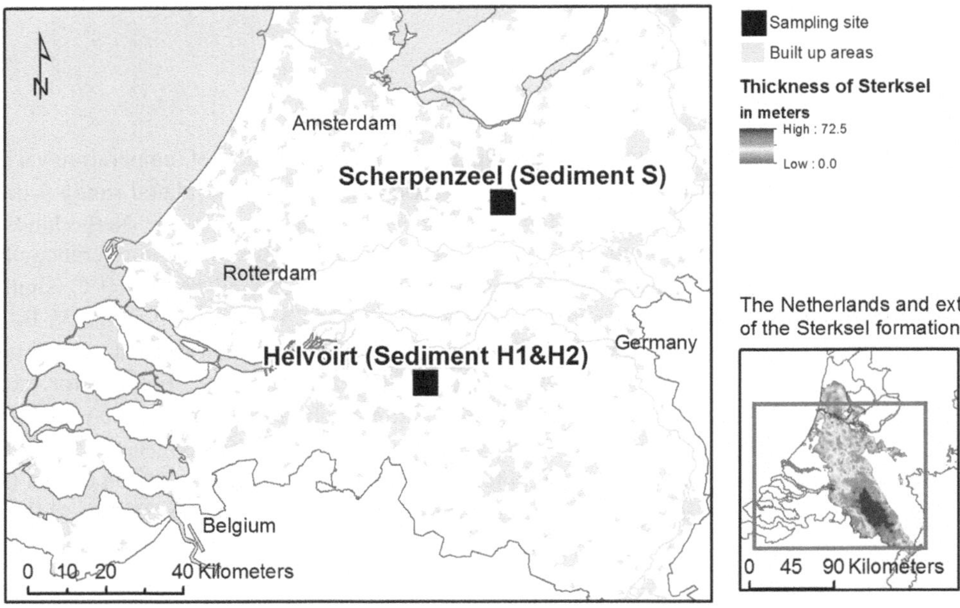

Figure 4.1 Sediment sampling locations and extent of Sterksel Formation (data courtesy of the REGIS geological database).

The sediment samples were collected using cores from boreholes drilled with the percussion drilling method. A minimum amount of drilling water (drinking water produced from the Sterksel aquifer sparged with N_2 gas to make it anoxic) was used to avoid contamination and keep sediments anoxic. Samples H1 and H2 were stored in a glovebox under nitrogen atmosphere. Sample S was stored in a second PVC liner filled with water sparged with N_2 gas and spiked with an anti-oxidant ($NaHSO_3$) in order to avoid permanent gas use in the glovebox.

A combination of geochemical analyses was used to characterize the sediments (Table 4.1, Table S1.1 in Appendix 1). The measured geochemical parameters were also used to calculate relevant reactive capacity properties: pyrite content, non-pyrite reactive iron (iron oxides) based on empirical relations derived for Dutch sediments (including the geological formations used here) reported by Griffioen *et al.* (2012).

Table 4.1 Geochemical and hydraulic characteristics of the sediments used in the column experiments.

Sample code and Location			H1: Helvoirt	H2: Helvoirt	S: Scherpen-zeel
Sampling depth (m-surface level)			44–46	32–34	34–36
Measured parameters	**Method**	**Unit**			
Reactive organic matter (LOI330)			0.24	0.15	0.06
Bulk organic matter (LOI550)[a]	Thermogravimetric analysis: Loss on ignition (LOI) at 330, 550 and 1000°C	% d.w.	0.76	0.48	0.14
Carbonate content (LOI1000)			6.00	0.54	0.74
Clay content	Laser grain size analyses (Fritch A22)		3.76	2.21	1.34
Median grain size (d_{50})		μm	187	244	310
S	Carbon Sulfur (CS) Combustion Analyzer (LECO)	% d.w.	0.1	0.03	0.05
Minerals present	X-Ray diffraction (XRD, Bruker D8 advance)		quartz, K-feldspar, albite, ferrian clinozoisite(C), muscovite (A & B), calcite, dolomite, ankerite (B)		
Cr_2O_3			0.037	0.028	0.001
Fe_2O_3	X-ray fluorescence (XRF)	%	1.36	1.02	0.59
Al_2O_3			4.04	4.55	3.36
As			5.7	2.9	3.7
Mo	$LiBO_2$ fusion followed by HNO_3 digestion and inductively coupled plasma – mass spectrometry (ICP-MS)	ppm	0.2	5.8	0.1
V			18	17	<8
Be			2.00	<1	<1
Calculated parameters	**Equation**	**Unit**			
Bulk density, ρ_b	Weight/volume	g cm^{-3}	1.94	1.94	2.12
Porosity, n	$n = (\rho_{bulk} - \rho_{H_2O}) / (\rho_{solid} - \rho_{H_2O})$		0.43	0.43	0.32
Pyrite[b]	$0.5 * S * M_{FeS_2} / M_S$	% d.w.	0.19	0.06	0.09
Reactive iron (iron oxides)[b]	$2 * M_{Fe} / M_{Fe_2O_3} * (Fe_2O_3 - (0.225 * Al_2O_3 - 0.91\%)) - 0.5 * S * M_{Fe} / M_S$		0.87	0.61	0.48

[a]corrected for structural water loss from clay according BOM = LOI550 – 0.7 × Clay% (Breeuwsma, 1987).
[b]based on Griffioen et al. (2012) with M_{FeS_2}, M_{Fe}, M_S, $M_{Fe_2O_3}$ being the molar masses of FeS_2, Fe, S, and Fe_2O_3, respectively (55.8, 32.1, 159.7), and Fe_2O_3 and Al_2O_3 the % d.w. of iron oxides and aluminum oxides in sediment determined by XRF, and S being the content of sulfur determined with the CS combustion analyzer.

4.2.2 Influent water collection and characterization

The boreholes used to collect the cores were completed as monitoring wells and used to collect groundwater that served as influent for the column experiment. The depth of the well screen corresponded within 5 m from the depth that the sediment was collected from. Groundwater was collected in a stainless steel 60 L pressure barrel using a submersible sampling pump at 20 m below the water table. Prior to sampling, the pressure barrel was first depressurized to 0.4 atm and then filled with pure N_2 gas to 1 bar pressure for three consecutive times. The monitoring well was purged at around 1 m^3/hour for 30 minutes prior to sampling while assuring that electrical conductivity (EC), pH, temperature, and dissolved oxygen (DO) were stable. Groundwater was filtered in-line with a 0.45 μm filter. The influent barrel was filled completely, thereby removing the complete gas filled head space with an adjustable pressure relieve valve. Collected influent was chemically characterized over the running time of the experiment using the analytical techniques discussed below. Results are summarized in Table 4.2 showing mean concentrations and standard deviation. The partial CO_2 pressure (P_{CO_2}), redox potential (pe), and charge balance were calculated with PHREEQC (Parkhurst & Appelo, 1999).

Table 4.2 Chemical characteristics of the influent water showing mean concentrations and standard deviations. Detection limits are shown in Table S1.2 in Appendix 1.

Parameter	Unit	Detection limit	Effluent H1	Effluent H2	Effluent S
pH	–		7.0 ± 0.1	6.9 ± 0.2	7.3 ± 0.2
EC(20°C)	μS cm^{-1}		240 ± 5	232 ± 10	202 ± 5
Alkalinity	meq L^{-1}		1.4 ± 0.1	1.3 ± 0.1	2.0 ± 0.2
P_{CO_2}	atm		0.010	0.010	0.006
pe	–		–0.4	–1.4	–1.0
Charge balance	%		–2.40%	3.30%	0.90%
Cl	mg L^{-1}	0.1	27.8 ± 0.8	27.6 ± 0.6	7.1 ± 0.2
SO_4		0.1	8.0 ± 0.3	7.9 ± 0.1	6.0 ± 0.1
NO_3		0.1	<0.25	<0.25	<0.25
NH_4		0.01	0.4 ± 0.1	0.5 ± 0.1	0.1 ± 0.01
DOC		0.2	2.3 ± 0.07	2.0 ± 0.06	0.6 ± 0.03
Na		0.006	12.0 ± 0.1	13.5 ± 0.9	5.6 ± 0.2 K
K		0.004	1.1 ± 0.02	1.6 ± 0.4	0.8 ± 0.1
Ca		0.02	28.2 ± 2.0	25.7 ± 2.0	29.9 ± 0.8 Mg
Mg		0.01	2.2 ± 0.2	2.3 ± 0.3	2.2 ± 0.1
Fe		0.001	0.3 ± 0.1	3.7 ± 3.2	0.3 ± 0.1
Mn		0.001	1.5 ± 0.01	0.9 ± 0.01	0.1 ± 0.01
Si		0.01	10.4 ± 0.1	10.8 ± 0.3	5.2 ± 0.9
tot-P	μg L^{-1}	5	12 ± 7	43 ± 8	65 ± 20
Al		3	<1	1 ± 0.5	<1
As		1	2 ± 0.2	6 ± 2	6 ± 5.8
B		4	4 ± 0.1	4 ± 1	1 ± 0.4
Cr		0.5	0.6 ± 1	1.0 ± 2	0.9 ± 1
F		20	70 ± 23	47 ± 15	64 ± 6
V		1	<1	<1	<1
Mo		1	1 ± 0.2	1 ± 0.1	2 ± 1
Be		0.02	<0.02	<0.02	<0.02

4.2.3 Experimental setup

Sediment cores were unpacked in a glove box under a N_2 atmosphere, mixed thoroughly for 30 minutes using stainless steel spatulas to obtain a homogeneous sample, and repacked in four 0.4 m long stainless steel cores (polished and HNO_3 washed (pacified) steel quality SS304) with an internal diameter of 0.066 m.

The cores were then placed in the experimental set-up shown in Figure 4.2 and kept at temperatures of 5, 11, 25, and 60°C. The entire setup was placed in a 11°C climate room, the 5°C core was additionally cooled with a circulation cooler, and the 25 and 60°C cores were heated with heating mats (Mechaheat custom made). Influent water was kept at a constant excess pressure of 1 atm (compared to atmospheric pressure) by replacing the water removed from the tank by a mixture of N_2/CO_2 gas. The CO_2 fraction was adjusted to the value under field conditions. A flux of 0.37, 0.37, and 0.33 mL/min was applied to sediments H1, H2, and S, respectively, resulting in a 1 day residence time. A pressure relieve valve maintained an excess pressure within the sediment cores of 1 atm. The length of the experiment was 24 days for each sediment resulting in 24 pore flushes. Dissolved oxygen (D.O.) and pH of the influent were continually logged with an in-line sensor (Mettler Toledo Easysense pH31 and $O_2$21).

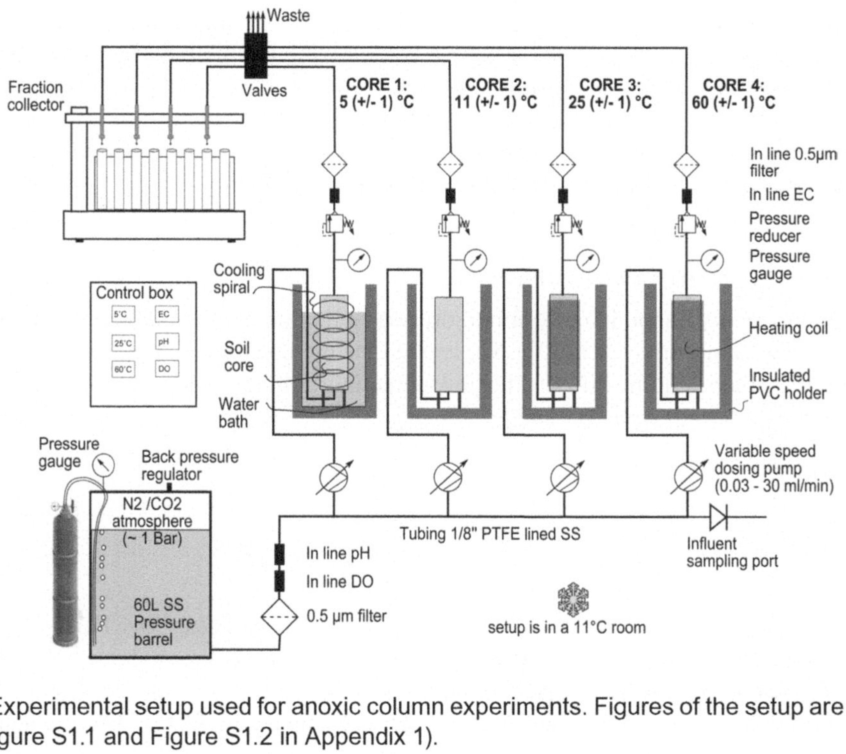

Figure 4.2 Experimental setup used for anoxic column experiments. Figures of the setup are shown in the Appendix (Figure S1.1 and Figure S1.2 in Appendix 1).

Steel cores were used because it was essential to keep the sediment anoxic and frequently used plastics like Teflon or Polypropylene are permeable to oxygen (Kjeldsen, 1993). Stainless steel quality SS304 (EN standard 1.4301) used here however contains Ni (8–11%) and Cr (18 and 20%) and this could thus interfere with the testing results. Blank testing of this steel using demineralized water indicated that only Ni and

Fe can leach in significant amounts of 7 μg L^{-1} and 80 μg L^{-1} (at both 25 and 60°C), Mo was observed just above the MDL (2 μg L^{-1} at both temperatures), whereas no significant effects were measured for Cr and V (Table S1.3 in Appendix 1).

4.2.4 Hydrochemical analyses

Alkalinity of in- and effluents was measured by titration with H_2SO_4 down to a pH of 4.3. Major cations, trace elements, and heavy metals were analyzed with inductively coupled plasma optical emission spectrometry (ICP-OES, Varian 720 ES-axial), anions with ion chromatography (Dionex DX-120 IC equipped with IonPac AS14 column), NH_4 by colorimetry (Labmedics Aquakem 200), and DOC was measured using high temperature catalytic oxidation (Shimadzu Model TOC-500A analyzer). Quality of the analyses was verified by calculating the electrical balance with PHREEQC.

4.2.5 Data analysis

In order to get a comprehensive overview of the temperature dependence in leaching behavior for the substances, we calculated the mean leaching concentration during the testing period at 5°C, 25°C, and 60°C and compared it to the reference of 11°C (which is the in-situ temperature). The significance of the difference in concentrations was tested at a 95% confidence levels with a Mann-Whitney U (MWU) test. This test is comparable to the more frequently used t-test but does not require a normal distribution. Statistical calculations were performed using the SciPy module (Anonymous, 2012). All calculations for redox speciations and saturation indices (SI, the logarithm of the ratio of ion activity product and solubility product) were carried out using PHREEQC (Parkhurst & Appelo, 1999).

4.2.6 Arsenic sorption isotherms

Column experiments are well suited to determine sorption isotherms and their thermodynamic parameters (Griffioen *et al.* 1992; Limousin *et al.* 2007). We applied the method derived by Jacobson *et al.* (1984) to derive the sorption isotherms at the different testing temperatures for arsenic sorption to iron oxides using the arsenic elution data of sediments H2 and S:

$$S(C) = S_0 - \frac{1}{m}\int_{c}^{c_0} (V(c) - V_0)dc \qquad (4.1)$$

where $S(C)$ represents arsenic bound to iron oxides present in the sediment (μg As/g Fe) in equilibrium with the dissolved concentration c (μg L^{-1}), $V(c)$ is the volume of effluent (mL) passed to reach concentration c (μg L^{-1}), V_0 is the pore water volume of the column (mL), m is the mass of iron oxides present in column (g), which is equal to the weight percentage (%) of reactive Fe shown in Table 4.1 multiplied by the total sediment weight and S_0 is the initial amount of arsenic present in the column. We note that the sorption isotherm was not calculated for sediment H1, because at 5°C and 11°C, As concentrations remained below detection limit. Equation (4.1) essentially represents a mass balance tracking the sorbed concentration starting at S_0 at the start of the experiment, which is then reduced with progressive flushing of the sediment. S_0 is estimated by integrating the area under the 60°C arsenic leaching curve, extrapolated to the point where all arsenic is leached out. In order to extrapolate the eluent curve, we used Eq (4.2) which is based on the observation that As versus the total flushed volume shows an exponentially decreasing pattern. Plotting Ln[As] versus the total flushed volume thus yields a straight line. The area under the line, representing the total leachable amount of arsenic per g of iron oxide can then be calculated with:

$$S_0 = \frac{-C_0}{am} \tag{4.2}$$

where C_0 and a are the intercept (µg L^{-1}) and slope (L^{-1}), respectively, obtained from a plot of Ln[C] versus the flushed volume (L). The supplementary information contains plots to further detail the application of Eqs. (4.1) and (4.2) to the eluent data. The derived sorption isotherms were fitted with both Freundlich (Eq 4.3) and Langmuir (Eq. 4.4) isotherms, commonly used for As sorption (Kanel *et al.* 2005; Banerjee *et al.* 2008):

$$S_F = K_F C^{1/n} \tag{4.3}$$

$$S_L = S_{\max} \frac{CK_L}{1 + K_L C} \tag{4.4}$$

where S_F and S_L are the mass of arsenic sorbed (µg g^{-1}) for the Freundlich and Langmuir isotherm, respectively, C is the equilibrium aqueous arsenic concentration (µg L^{-1}), K_F is the Freundlich constant indicative of the adsorption capacity (L g^{-1}), and n is a fitting constant (dimensionless) indicative of the sorption intensity, K_L is the Langmuir constant (L g^{-1}), and $S_{\max}$ is the maximum sorption capacity (µg g^{-1}). The Freundlich parameters $1/n$ and log[K_F] are equal to the slope and intercept of a regression line of C versus S on a double logarithmic scale (Limousin *et al.* 2007). The Langmuir parameters were found by minimizing the sum of the squared errors (SSE) between observed S_{obs} and calculated S_L.

4.2.7 Sorption thermodynamics

Next, the impact of temperature on the sorption process was determined by calculating the change in standard Gibbs free energy (ΔG^0, kJ mol^{-1}), entropy (ΔS, kJ mol^{-1} T^{-1}), and the reaction enthalpy (ΔH, kJ mol^{-1}) for the sorption reaction using the following equations (Stumm & Morgan, 1996):

$$\Delta G^0 = -RT \ln K_d \tag{4.5}$$

$$\ln K_d = \frac{\Delta S}{R} - \frac{\Delta H}{RT} \tag{4.6}$$

where K_d is the distribution coefficient at infinite dilution, R the gas constant (8.314 J K^{-1} mol^{-1}), and T the absolute temperature (K). The entropy change and enthalpy were calculated from the intercept and slope of a van't Hoff plot (Ln[K_d] versus T^{-1}) where K_d is derived from the slope of the Freundlich ($K_{d,F}$ in Eq 4.7) and Langmuir ($K_{d,L}$ in Eq 4.8) isotherms at infinite dilution (or zero concentration), given by:

$$K_{d,F} = K_F/n \tag{4.7}$$

$$K_{d,L} = K_L S_{\max} \tag{4.8}$$

4.3 RESULTS AND DISCUSSION

4.3.1 General patterns

Table 4.3 shows which elements were significantly influenced in all three experiments based on the MWU test. Table 4.3 also shows whether the elements were identified in the sediments by XRF and HNO_3/ICP-MS. This comparison shows that most heavy metals and trace elements that were not mobilized by increased temperatures were actually present in the sediment. Figure 4.3 presents the effluent graphs from sediment S for species showing a statistically significant temperature variation (As, K, Si), and for parameters and elements which showed considerable change from the influent concentration (pH, Alkalinity, Ca, and Mg). The effluent graphs for sediments A and B are comparable to those of sediment S and are included in the supplementary information.

Table 4.3 Summary of leaching behavior of the 5°C, 25°C, and 60°C columns compared to the reference column of 11°C during 1 day residence time and presence of substances in sediment (detection limit of the geochemical analyses are shown in the supplementary info).

Leaching behavior	Geochemical	Temperature level		
		5°C	25°C	60°C
Substances significantly thermally influenced ($p < 0.05$) in all three experiments,	Substance present in sediment		As	As, B, Mo, V, K, Si, DOC, P
	Not analyzed			F, pH
Leaching behavior not significantly influenced by temperature in all three experiments	Substance present in sediment	Na, Ca, Mg, Sr, Fe, Mn, Al, Ba, Be, Co, Cr, Cu, Ni, Pb, Zn, Eu, Ho, Sb, Sc, Yb		
	Substance below detection limit in sediment	Ag, Cd		
	Not analyzed	Alkalinity, SO_4, Br, Cl, Tl		
Substance below detection limit in reference and testing temperature	Substance present in sediment	Ga, La, Th		
	Substance below detection limit in sediment	Bi, Se		

A temperature reduction from 11°C to 5°C did not cause significant differences in mean effluent concentrations for any substance in all three tested sediments. However, several substances showed significant differences in one or two sediments. For example, K showed a decrease at 5°C in all three sediments but the difference was only significant in sediment S. The 25°C effluent showed a significant increase in As concentrations for all three sediments. In sediments H2 and S, As concentrations exceeded the drinking water standard of 10 $\mu g\ L^{-1}$ from the European Union drinking water directive (Anonymous, 1998). The increase in As observed at 25°C in sediment A is less pronounced than that in sediments H2 and S, but still significant at a 95% confidence level. A significant increase in K concentration was observed at 25°C in sediment S but not for the other two sediments.

At 60°C, many substances and pH showed significantly increased concentrations: As, F, DOC, P, K, Si, B, Mo, and V. All 60°C effluents exceeded the drinking water standard for As. Concentrations of all these substances increased especially during the initial pore flushes, and leveled off with increasing pore

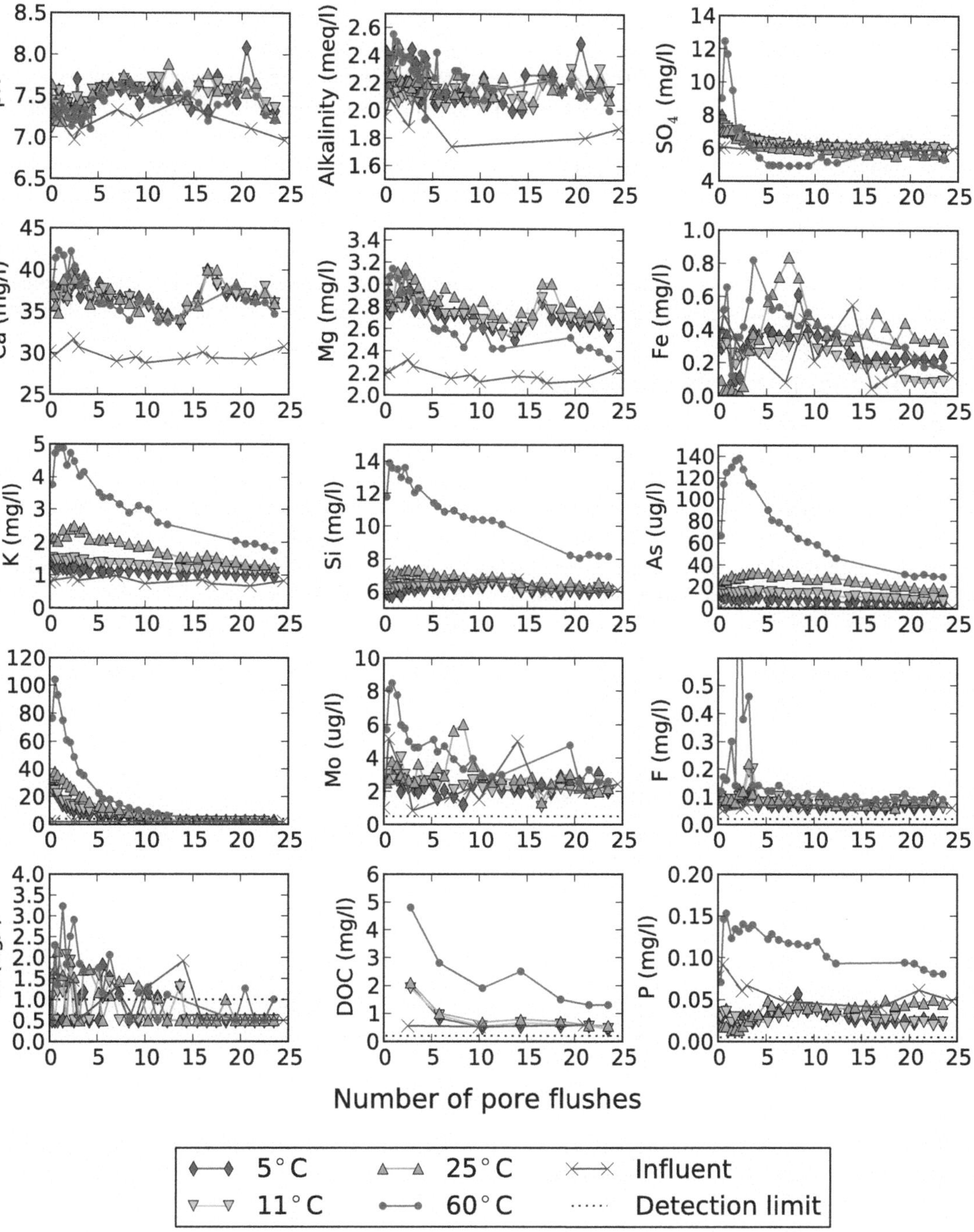

Figure 4.3 Selection of effluent graphs for the main cations and anions, trace elements, and DOC for sediment S. Results for the 60°C column between pore flushes 13 and 19 are not shown because of sampling errors.

flushes. These patterns are clearly visible in Figure 4.3 which also shows that concentrations of substances like DOC, Si, and K decreased less rapidly than those of As and Mo. In the 60°C effluents of sediments H1 and S, elevated concentrations of Be and Cr were measured (graphs are included in the supplementary information). The content of Cr_2O_3 in sediments H1 and H2 (respectively 0.037% and 0.028%) was also considerably higher than in sediment S (0.001%) indicating that if Cr is present in the sediment it may become mobilized in response to a temperature increase. Although these increases were significant, the concentrations ranged between 1 and 2 μg L^{-1} at 60°C (compared to 0.6 to 1 μg L^{-1} for the reference) and were well below the drinking water standard of 50 μg L^{-1}. For Be the difference between the different sediments was less pronounced (2 ppm, <1 ppm, and <1 ppm in sediments H1, H2, and S, respectively). In the following we discuss the main hydrogeochemical processes triggered by the temperature changes, distinguishing dissolution of silicate minerals, mineralization of organic carbon, and trace metal desorption.

4.3.2 Silicate minerals dissolution

In both the 25°C and 60°C effluents, a correlated increase in K and Si concentrations was observed ($R^2 > 0.8$, Figure 4.4) indicating dissolution or transformation of a silicate mineral such as muscovite or K-feldspar (both identified by XRD). In order to gain insight into which minerals may dissolve upon a temperature increase, Table 4.4 presents the SI-values for various silicate minerals showing that at 5 to 25°C, influent was under saturated with respect to K-feldspar, while at 60°C influent became under saturated with respect to quartz, kaolonite, and gibbsite. The Si:K ratio of 2.5 observed in the 60°C effluent is between the stoichiometric ratios for incongruent K-feldspar dissolution with formation of kaolinite (Si:K = 2:1, Eq (4.9)), and congruent dissolution (Eq (4.10) Si:K 3:1) (Appelo & Postma, 2005):

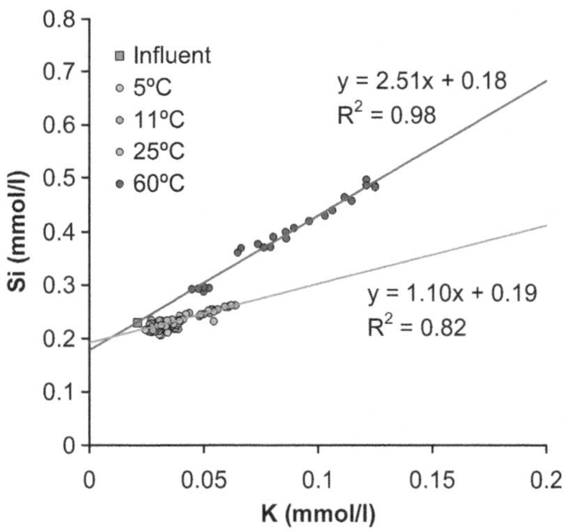

Figure 4.4 Relationship between Si and K molar concentrations in sediment S effluent. Regression lines are drawn for the 25°C and 60°C data of sediment S.

$$2KAlSi_3O_8 + 2H^+ + 9H_2O \rightarrow Si_2Al_2O_5(OH)_4 + 2K^+ + 4H_4SiO_4^0 \quad (4.9)$$

$$KAlSi_3O_8 + 8H_2O \rightarrow K^+ + Al(OH)_4^- + 3H_4SiO_4^0 \quad (4.10)$$

Table 4.4 Calculated saturation indices for various silicate minerals and gibbsite for influent S. Calculations were carried out with PHREEQC (Parkhurst & Appelo, 1999). Minerals for which water is oversaturated are highlighted in bold.

Temperature (°C)	Sample	Quartz (SiO_2)	K-feldspar ($KAlSi_3O_8$)	K-Mica ($KAl_3Si_3O_{10}(OH)_2$)	Kaolinite ($Al_2Si_2O_5(OH)_4$)	Gibbsite ($Al(OH)_3$)
5	Influent	**0.3**	−1.6	**6.2**	**3.8**	**1.0**
	Effluent 1 pore flush	**0.3**	−1.4	**6.3**	**3.5**	**0.9**
11	Influent	**0.2**	−1.6	**6.8**	**4.1**	**1.2**
	Effluent 1 pore flush	**0.2**	−0.9	**8.1**	**4.7**	**1.6**
25	Influent	0.0	−2.2	**5.8**	**3.2**	**1.0**
	Effluent 1 pore flush	0.0	−1.8	**6.0**	**3.0**	**0.9**
60	Influent	−0.4	−4.6	**0.9**	−0.4	−0.4
	Effluent 1 pore flush	−0.1	−3.1	**2.3**	**0.1**	−0.4

This suggests that (partial in)-congruent dissolution of K-feldspar could explain the elevated Si and K concentrations. This corresponds with the findings of Holm *et al.* (1987) who related elevated K concentrations during high temperature energy storage (90–110°C) to K-feldspar weathering. It is remarkable that the Si:K ratio of 1.1 at 25°C was much lower than at 60°C. A possible explanation for this is that at 25°C influent was saturated with respect to quartz ($SI_{quartz} = 0.0$) while at 60°C it was sub-saturated ($SI_{quartz} = -0.4$). This could imply that at 25°C, K-feldspar dissolved incongruently with formation (amorphous) quartz, where the secondary precipitating minerals causes removal of Si and lowering the Si:K ratio. This process was also observed by Wollast (1967) who described that Si release from K-feldspar dissolution at room temperature is controlled by the reaction with $Al(OH)_3$ resulting in the precipitation of hydrated silica. In addition, Wollast (1967) described that at room temperature the initial step K-feldspar weathering was incongruent involving the exchange of fixed K^+ with aqueous H^+, while at 200°C dissolution was initially congruent with formation secondary minerals (Eqs (4.9) and (4.10)).

4.3.3 Dissolved organic carbon mobilization

Dissolved organic carbon (DOC) increased at 60°C in all sediments. Two mechanisms could be responsible for these increases: (1) desorption of organic acids from iron oxides (Evanko & Dzombak, 1998; Filius *et al.* 2000) and (2) microbial respiration of SOM (Brons *et al.* 1991). A key difference between these two processes is that DOC involved in sorption generally comprises large fulvic or humic acids (Filius *et al.* 2000) while respiration of SOM is a stepwise transformation of SOM to volatile fatty acids such as acetate which are then consumed by sulfate reducing bacteria or methanogenic *Archaea* generating inorganic carbon but not necessarily DOC (Muyzer & Stams, 2008). In the second process DOC would therefore accumulate with increasing temperature only if the rate of DOC oxidation is lower than that of DOC formation. Robador *et al.* (2011) and Weston and Joye (2005) observed such a decoupling between the different steps when marine sediments were incubated above their ambient temperature, leading to an accumulation of DOC. Brons *et al.* (1991) however concluded that the production of DOC at temperatures above 30°C becomes abiotic and should be viewed as pure chemical mineralization of SOM.

The relative contribution of the two different processes leading to DOC accumulation was hard to quantify. However, a qualitative indication could be obtained since different elution profiles can be expected: desorption will generate a peak in DOC which will gradually decrease towards the influent concentration with continuing flushing, whereas SOM-derived DOC would yield a constant DOC concentration with progressive flushing (under equal residence time) assuming ample SOM in the sediment with constant reactivity. The elution curves for sediments H1 and especially H2 (Figures S1.3 and S1.4 in the Appendix) were characterized by a short peak (<7 pore flushes) followed by constant increase relative to the influent DOC. In sediment S, a gradual decrease in DOC was observed up to 20 pore flushes after which DOC stabilizes just above the influent DOC concentration. This suggests that in sediments H1 and H2, the DOC increase was mostly derived from SOM, whereas in sediment S it was mostly derived from desorption. This qualitative observation is in agreement with the geochemical analyses: reactive SOM in sediment S was much smaller than that in sediments H1 and H2, whereas iron oxide content was also lower in sediment S than in sediments H1 and H2 but to a considerable smaller extent than for the SOM difference.

4.3.4 Mobilization of arsenic and other trace compounds

Various mechanisms responsible for arsenic mobilization in aquifers are reported in the literature (Smedley & Kinniburgh, 2002, Stuyfzand *et al.* 2006) which can be divided in three main groups: (1) oxidation of pyrite, (2) reduction of Fe(III) minerals, or (3) desorption from reactive surfaces such as Fe, Al, and Mn-oxides.

Oxidation of As containing pyrite is not likely as the influent DO and NO_3 were consistently below the detection limit during the experiment (DO < 0.05 mg L^{-1} and NO_3 < 0.1 mg L^{-1}) and the decreased SO_4 concentrations in the 60°C effluent beyond the first three pore flushes point to the occurrence of sulfate-reduction. The elevated SO_4 concentrations during the first pore flushes in the 60°C effluent were unlikely the result of pyrite oxidation because of the aforementioned arguments and given the similarities between the four columns. Iron oxides are reported to be able to sorb SO_4 (Zhang & Sparks, 1990; Wijnja & Schulthess, 2002) but we did not find any literature on the temperature dependence of sorption. The initial SO_4 peak in the 60°C effluent may be explained by exothermic sorption, thus releasing SO_4 under increased temperature.

Microbiologically mediated reductive dissolution of iron (hydr)oxides can be stimulated by a temperature increase as demonstrated by Jesußek *et al.* (2012). In our observation data, however, there was no clear Fe concentration increase, suggesting that reductive dissolution is not the main source of As. It is however possible that some of the reductively dissolved Fe(II) reacts with HS^- formed during sulfate reduction to precipitate as an iron sulfide mineral, which could then sequester part of the released As, a process utilized for As remediation (Beaulieu & Ramirez, 2013). This process could thus only contribute at 60°C where sulfate reduction is occurring, masking the effect of reductive iron dissolution.

The third general explanation, desorption of arsenic, is widely described for both iron oxides (Smedley & Kinniburgh, 2002; Dixit & Hering, 2003; Maji *et al.* 2007; Kersten & Vlasova, 2009a) and expected to be the controlling mechanism for sediment S (which yielded the highest As concentrations) and to a lesser degree in sediment H2. In these sediments, a decrease in As concentration was observed with progressing pore flushes which may be explained by desorption of As until a new equilibrium is reached. The clearest decrease in As concentrations was observed in sediment S, while in sediment H2 a sharp initial rise in concentration was observed followed by a decrease towards a relatively stable level. The 25°C and 60°C leaching concentrations of As in sediment H1 remained relatively constant with progressive flushing. The relatively stable level in As concentrations in the late time data of sediment H2 and that of sediment H1, suggests that dissolution from an As-containing mineral is contributing to the enhanced As concentrations in these sediments. It was however not possible to obtain a quantitative estimate of the relative contribution of both processes, this would require additional data that can be used to constrain the progression of reductive dissolution of iron oxides coupled to organic carbon mineralization such as NH_4 or CH_4. Also experiments with a longer and increasing residence time would be required to distinguish kinetically controlled dissolution from desorption which is a relatively fast process (Appelo & Postma, 2005).

Desorption from iron oxides can be driven by various mechanisms of which the most relevant is the temperature induced change in sorption equilibrium constant as shown for example by Kersten *et al.* (2009a) in laboratory experiments with synthetic goethite. The release of As could be further stimulated by anions which compete with As for sorption sites (van Halem *et al.* 2012) such as PO_4 or DOC produced by SOM mineralization and H_4SiO_4 produced by silicate mineral dissolution.

Note that other elements mobilized at 60°C, such as Mo, V, and B, all occur as oxyanions like As and sorb to iron(hydr)oxides (Dzombak & Morel, 1990). This similarity in chemical behavior, may suggest that indeed desorption from, or reductive dissolution of, iron oxides was also responsible for the mobilization for these trace elements.

4.3.5 Arsenic sorption isotherms

In order to investigate the temperature dependence of sorption in a more quantitative way, the eluent data was used to construct Freundlich and Langmuir isotherms for sediments H2 and S using Eqs 4.1 to 4.4 (Figure 4.5 and Table 4.5) where it is assumed that the occurrence of As in the effluent was caused by sorption only. The comparison of experimental data and fitted Langmuir and Freundlich isotherms shows a good agreement between model and data. The Langmuir model is expected to provide a more robust description of the sorption process because it can be expected that the sorption of As was maximized by the number of sorption sites. The data in Table 4.5 also show that the sorption isotherms for 5, 11, and 25°C only cover a narrow concentration spectrum. The derived sorption isotherms are therefore considered accurate only within the range of measured aqueous As concentrations. In order to derive a complete isotherm representative for a large concentration interval, the sorbent should first be completely covered with As which would require a different experimental setup. The sorption isotherms are typical L curve types as described by Sposito (2008). This type of curve points to a very high affinity of the sorbent for the solute, and is generally due to an inner sphere surface complexation reaction (Sposito, 2008). This corresponds to results of spectroscopic investigations on As sorption on iron oxides by (Goldberg & Johnston, 2001) showing that inner sphere surface complexation is the driving sorption mechanism.

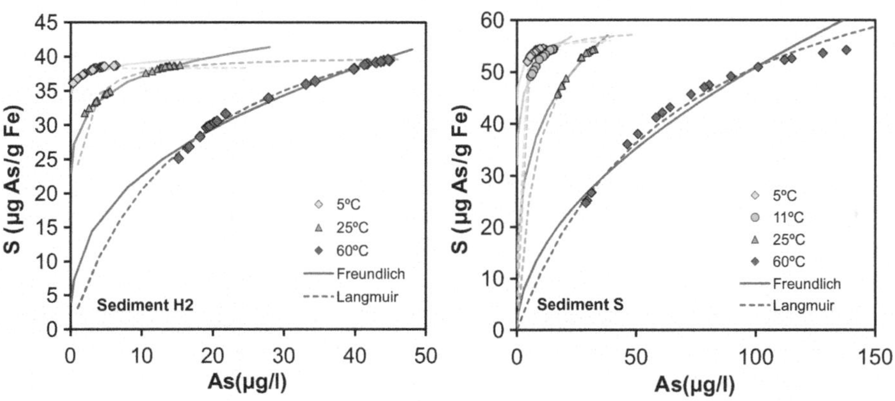

Figure 4.5 Sorption isotherms for arsenic using data from sediments H2 and S. Markers are for observed aqueous concentrations and calculated sorbed masses. The fitted Freundlich and Langmuir isotherms are shown with solid and dashed lines, respectively.

Table 4.5 Freundlich and Langmuir sorption parameters for arsenic sorption on iron oxides in sediments B and C.

	Sediment H2			Sediment S			
	5°C	25°C	60°C	5°C	11°C	25°C	60°C
Freundlich parameters							
K_F (L g^{-1})	36.8	29.3	9.5	48.7	40.9	21.2	4.4
1/*n* ( )	0.03	0.10	0.38	0.05	0.10	0.27	0.53
Langmuir parameters							
K_L (L g^{-1})	37.7	1.5	0.06	2.7	0.9	0.1	0.015
S_{max} (μg As/g $Fe_{reactive}$)	38.4	40.2	54.3	56.5	58.4	69.7	85.0

4.3.6 Temperature influence

The van't Hoff plots for As sorption based on the two sorption models are shown in Figure 4.6 and the thermodynamic parameters based on Eqs 4.5 and 4.6 are shown in Table 4.6. The negative ΔH indicates As sorption was exothermic and the affinity of the sorbent for As decreases with increasing temperature. There is some variation in enthalpies calculated for the different sediments and models but overall the enthalpies for different sediments and models agree well.

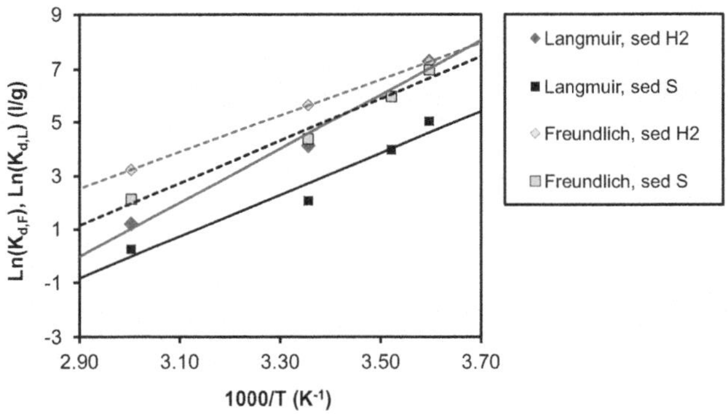

Figure 4.6 Van't Hoff plots: Ln[*K*] versus 1000 T^{-1} for experimental data of arsenic from sediment H2 and S using Freundlich and Langmuir isotherms (solid symbols).

In order to compare the thermodynamic data of As sorption to iron oxides derived here, Table 4.7 presents a compilation of data derived from the literature confirming that As sorption is generally an exothermic process. However, the data for akageneite (a type of iron oxide) shows endothermic sorption behavior. The spread in ΔH values for the iron oxides with As is quite large: –161.8 to –26 kJ mol^{-1}, with our ΔH values being at the higher (negative) end. The large range in ΔH probably relates to (1) sedimentary differences, the type of iron oxide sorbent (degree of crystallinity or aging), and (2) the method applied to calculate ΔH: for example, Kanel *et al.* (2005) calculated the enthalpy directly from K_F and K_L and not from the K_d which is representative for infinite dilution, while others use a surface complexation model (e.g., Kersten & Vlasova (2009a)). Furthermore, for the studies using K_d, it was unclear how this K_d was calculated. In most studies, it is stated that K_d represented 'equilibrium conditions' (Maji *et al.* 2007; Hong *et al.* 2011) without describing how this was determined.

Table 4.6 Thermodynamic parameters: Gibbs energy, change in entropy, and change in enthalpy for arsenic sorption and R^2 between Ln[K] and T^{-1} on sediments H2 and S.

Sorption model	Sediment H2			Sediment S			
	5°C	25°C	60°C	5°C	11°C	25°C	60°C
Freundlich							
$\mathrm{Ln}[K_{d,F}]$	3.6	3.4	2.3	3.9	3.7	3.1	1.5
ΔG^0 (kJ mol^{-1})	−8.3	−8.4	−6.2	−9.0	−8.8	−7.6	−4.1
ΔS^0 (J K^{-1} mol^{-1})		−17.2±0.2			−21.8±2.8		
ΔH (kJ mol^{-1})		−56.4±0.4			−65.8±6.8		
R^2		1.00			0.98		
Langmuir							
$\mathrm{Ln}[K_{d,L}]$	3.6	0.4	−2.8	1.0	−0.1	−2.2	−4.2
ΔG^0 (kJ mol^{-1})	−16.8	−10.2	−3.3	−11.6	−9.4	−5.1	−0.7
ΔS^0 (J K^{-1} mol^{-1})		−29.3±4.4			−23.5±4.3		
ΔH (kJ mol^{-1})		−84.0±11.1			−64.8±10.7		
R^2		0.98			0.95		

Table 4.7 Reported reaction enthalpies for arsenic sorption.

Main minerals	Temperature range	ΔH (kJ mol^{-1})	Comment	Reference
This study				
Natural sediments	5,11, 25, 60°C	−56.4 to −84.0		
Other studies				
Waste containing mainly Fe-oxides	20,25,30°C	−161.8	ΔH based on K_d	(Negrea *et al.* 2010)
Laterite soil with mixture of Fe and Al oxides	25, 35, 45°C	−109.5	ΔH based on K_d	(Maji *et al.* 2007)
Zero valent Fe	25, 35, 45°C	−11.9	ΔH based on K from Freundlich isotherm	(Kanel *et al.* 2005) Values were corrected from original published data to correct for erroneous value for gas constant used.
	25, 35, 45°C	−112.41	ΔH based on K from Langmuir isotherm	
Goethite	10,25,50,60°C	−26.0 ± 5	Sorption of As on goethite	(Kersten & Vlasova, 2009a)
	10,25,50,60°C	−27.6 ± 5		
Fe-Al Binary Oxide	25,40,60°C	−12.64	ΔH based on K_d	(Hong *et al.* 2011)
Akaganeite	20,30,40°C	+8.87	ΔH based on K_d	(Banerjee *et al.* 2008)

Sorption isotherms provide a somewhat simplified description of the actual sorption process, which involves complexation of species at the oxide-water interface and strongly depend on the aqueous composition. Enthalpy data derived from sorption isotherms can only be used for similar groundwater compositions and iron oxides types. A more robust approach simulating sorption is provided by surface complexation models, which describe the intrinsic sorption reactions and can deal with variations in background chemistry and pH (Appelo & Postma, 2005).

4.3.7 Environmental implications

A number of environmental implications can be drawn from our data that are relevant for the groundwater's usability as raw water source in aquifers used for geothermal energy. First, various toxic substances are shown to be increasingly mobile under elevated temperatures (As, Mo, and V in sediments H1, H2, and S, and Be and Cr in sediments H1 and H2). If sorption is indeed the key process, the mechanism is probably reversible; when the temperature lowers resorption of these elements to iron oxides can be expected. However, there is a catch: field data show that ATES systems have a significant effect on water quality simply due to the mixing of different hydrochemical groundwater types in stratified aquifers (Chapter 3). Therefore, circulation of groundwater by ATES, will distribute heated and As enriched groundwater from layers rich in iron oxide to aquifer layers having lower contents of iron oxides having consequently lower capacity for As resorption during cooling. Even though the thermal impacts of the ATES system might dissipate when the system is de-commissioned, the effects of mobilization and re-distribution of As and other elements will be irreversible. If As is released by reductive dissolution of iron oxides, the process is irreversible and can proceed over a much longer time frame than desorption, potentially resulting in even higher As concentrations. However, if both iron and sulfate reduction occur, iron sulfide minerals may form which can effectively sequester As (Beaulieu & Ramirez, 2013).

Here, we used anoxic sediments where As is present as arsenite (based on PHREEQC speciation calculations) and it is interesting to speculate what may happen under oxic groundwater conditions where arsenic is present as arsenate. Two processes can impact on arsenic mobility: Firstly, (Jesußek *et al.* 2012) showed that in oxic sediments, a temperature rise can increase DOC concentrations and create more reducing conditions by faster reduction rates. Smedley and Kinniburgh (2002) showed that in many environments, this can trigger the mobilization of As. Secondly, it is likely that sorption of As(V) is also an exothermic reaction because both As(V) and As(III) form inner-sphere surface complexes with iron oxides (Goldberg & Johnston, 2001) implying that also in oxic aquifers used for heat storage, As may be mobilized.

A second environmental aspect is the increased DOC concentration in water which can give water a yellowish color. This could cause the requirement of a discoloration treatment step of the water abstracted for drinking water supply (Wallage *et al.* 2006). In addition, the mobilization of organic carbon leads to a reduction in SOM which in turn is a very important factor in sorption and immobilization of organic micro pollutants (Appelo & Postma, 2005). Furthermore, the humic substances as part of DOC can form complexes with trace elements, causing them to remain mobile and not to become resorbed with decreasing temperature outside the direct zone of influence of the ATES system (Appelo & Postma, 2005).

4.4 CONCLUSIONS

Our experiments showed that water quality was not affected when anoxic aquifer sediments were subjected to lower (5°C) than in-situ temperature (11°C). At the currently allowed maximum hot well temperature (25°C), As concentrations were significantly increased. At 60°C, significant effects were observed on the pH and DOC, P, K, Si, As, Mo, V, B, and F concentrations. Consequently, these elements should be considered for water quality monitoring programs of ATES projects. Arsenic exceeded the drinking water guideline of 10 μg L^{-1} in two of the three tested sediments at 25°C and in all sediments at 60°C. No consistent temperature effects were observed trace elements like Ba, Co, Cu, Ni, Pb, Zn, Eu, Ho, Sb, Sc, Yb, Ga, La, and Th, all of which were present in the sediment. These results highlight the need for careful evaluation of hydrogeochemical conditions when situating ATES systems in aquifers used for drinking water production. The temperature induced chemical effects are believed to be caused by various processes: The (in)congruent dissolution of silicates (K-feldspar) was most likely responsible for the increased levels of K and Si. The increasing DOC at 60°C may be explained by a combination of desorption and SOM mineralization. Desorption from, and potentially reductive dissolution of iron oxides is believed to be responsible for increasing concentrations of As, B, Mo, V, and P.

Chapter 5

Temperature-induced impacts on redox processes and microbial communities

ABSTRACT

In this chapter, we report the results of laboratory experiments on the effect of temperature variations (5°C to 80°C) on redox processes and associated microbial communities in anoxic unconsolidated sediments. Both hydrochemical and microbiological data showed that a temperature increase from 11°C (in-situ) to 25°C caused a shift from iron-reducing to sulfate-reducing and methanogenic conditions in a sediment containing reactive iron. Bioenergetic calculations could explain this shift. A further temperature increase (>45°C) resulted in the emergence of a thermophilic microbial community specialized in fermentation and sulfate-reduction. Two distinct maxima in sulfate-reduction rates, of similar order of magnitude (5×10^{-10} M s^{-1}), were observed at 40°C and 70°C. Thermophilic sulfate-reduction, however, had a higher activation energy (100–160 kJ mol^{-1}) than mesophilic sulfate-reduction (30–60 kJ mol^{-1}), which is probably due to a trade-off between enzyme stability and activity with thermostable enzymes being less efficient catalysts that require higher activation energies. These results reveal that while sulfate-reducing functionality can withstand a substantial temperature rise, other key biochemical processes appear more temperature sensitive.

5.1 INTRODUCTION

Previous research on shallow geothermal energy (SGE) systems, and aquifer thermal energy storage (ATES) revealed that groundwater quality is influenced by mixing of different water types (Chapter 2), temperature changes (Holm *et al.* 1987; Perlinger *et al.* 1987) or intrusion of oxygen, or degassing (Palmer & Cherry, 1984; Jenne *et al.* 1992). Most of the earlier research focused on operational aspects in relation to inorganic chemistry, such as well-clogging due to mineral precipitation occurring in high temperature systems (>25°C) (Palmer & Cherry, 1984; Holm *et al.* 1987; Perlinger *et al.* 1987; Griffioen & Appelo, 1993). In addition, we found in Chapter 4 that the mobility of arsenic and other oxyanions was significantly enhanced with a temperature increase, which was attributed to either desorption from, or reductive dissolution of iron oxides.

Reprinted with permission from Bonte M., Roling W. F. M., Zaura E., van der Wielen P. W. J. J., Stuyfzand P. J. and van Breukelen B. M. (2013). Impacts of shallow geothermal energy production on redox processes and microbial communities. Environmental Science and Technology, ***47****(24), 14476–14484.*

As discussed in Chapter 2, redox reactions are expected to be particularly sensitive to temperature changes. This expectation is based on research dealing with the impacts of relatively modest seasonal temperature changes which was shown to (i) strongly accelerate pyrite oxidation in a deep well injection experiment (Stuyfzand, 1998; Prommer & Stuyfzand, 2005), and (ii) be of great influence on mineralization of organic matter at river bank filtration sites (Greskowiak *et al.* 2006; Sharma *et al.* 2012). In addition, Jesußek *et al.* (2012) reported a shift in redox conditions, with oxic conditions towards nitrate- and iron-reducing conditions at 25 and 40°C and sulfate reducing conditions at 70°C. Redox processes in aquifers are often the last step of the microbial metabolism, involving hydrolysis and fermentation of organic matter, followed by oxidation by electron acceptors such as nitrate, ferric iron or sulfate, implying that the temperature-induced hydrochemical changes are related to the changes in the microbial community. Brielmann *et al.* (2009, 2011) showed that the microbial community was significantly influenced by changing temperature, both near an ATES system, and in laboratory experiments but did not describe the hydrochemical changes associated with this.

None of these previous studies looked in an integrated way at the temperature impacts on both microbial functioning and associated hydrochemical changes. In order to fill up this void, we used the column setup described in the previous chapter to study the influence of temperature (5–80°C) on the kinetics and competition of redox reactions, and associated changes in microbial communities. The novelty of this study is that we combined conventional hydrochemical analyses to determine the predominant redox processes, sulfate reduction rates as a proxy for microbial activity, and 454-pyrosequencing to map the changes in microbial community. We derived activation energies for sulfate-reduction, which was the dominant redox reaction in the investigated sediments, providing a comprehensive quantitative description of the temperature dependence, which can be incorporated in hydrochemical models dealing with temperature changes due to ATES.

5.2 MATERIALS AND METHODS

5.2.1 Sediment and groundwater collection

Sediments and groundwater used in the column experiments were collected at two locations from an aquifer intensively used for both drinking water production and ATES in the Netherlands: Helvoirt (sample H2, derived from 32 and 34 m-surface level (m-SL)) and Scherpenzeel (sample S, 34 and 36 m-SL). Note that sediment sample H1, described in the previous chapter was not used in the experiments described here. The unconsolidated sandy aquifer is part of the fluvial Sterksel formation ranging in thickness between 5 and 75 m, which was deposited during the Early to Middle Pleistocene (Westerhoff, 2003). Sediment cores were collected with the percussion drilling method and transported in an ice box filled with N_2 gas to assure cores remained anoxic. Geochemical analyses revealed that samples consisted mainly of quartz sand (>90%), with minor fractions of K-feldspar and clay (details on geochemical analyses given in Chapter 4). Reactive phases included organic matter (0.15 and 0.06% for sediments H and S, respectively), carbonates (0.5 and 0.7%), pyrite (0.06 and 0.09%), and iron-oxides (0.6 and 0.5%).

Influent for the column experiment was derived from groundwater abstracted from monitoring wells constructed in the boreholes for sediment sampling and collected in a stainless steel 60 L pressure barrel using a submersible sampling pump at 20 m below the water table. Groundwater in this aquifer is anoxic, lacking O_2 and NO_3, with SO_4 concentrations decreasing with depth (Figure 3.4) which may be due to sulfate-reduction or a change in historic input of SO_4 and NO_3 with groundwater recharge. Iron concentrations are 6.5 mg L^{-1} and 0.3 mg L^{-1} for Helvoirt and Scherpenzeel, respectively. At both sites, CH_4 concentrations in the aquifer at the sampling depth were generally <0.01 mg L^{-1}.

5.2.2 Experimental setup

The cores were unpacked in the laboratory under a N_2 atmosphere, mixed thoroughly for 30 minutes using stainless steel spatulas to obtain a homogeneous sample, and repacked in four 0.4 m long stainless steel cores with an internal diameter of 0.066 m. The cores were then maintained at temperatures of 5°C (representing cold storage), 11°C (ambient temperature), 25°C (maximum allowed regular ATES), and 60°C (high temperature ATES).

At the start of the experiments, the sediment cores were first flushed for 25 days with a residence time of 1 day allowing the microbiological community to acclimatize to the testing temperatures of 5, 11, 25 and 60°C. Data obtained during this period is described in the previous chapter. Following this period, the residence time of water in the sediments was stepwise increased to 30 days, to evaluate the occurrence of redox processes by measuring the concentrations of major redox species such as SO_4, Fe, DOC, and CH_4 as function of residence time. During each residence time step, water was stagnant in the columns until the required residence period was achieved. Following this, water in the column was sampled for analysis and subsequently flushed with influent for 6 pore volumes in two days, resetting the aqueous concentrations in the column. This experiment (called hereafter: the increasing residence time (IRT) experiment) was carried out for both sediments H and S.

In order to gain a more detailed insight into the temperature dependence of the redox kinetics, sediment S was subsequently subjected to an experiment where the temperature was increased from 5 to 80°C in steps of 5°C, each with a constant 5 day residence time (temperature-ramping (TR) experiment). The resulting data sets were used to calculate the sulfate-reduction rates (SRRs) at different temperatures and activation energies. A detailed sampling schedule is shown in the appendix to this chapter, Figure S2.1 in Appendix 2.

5.2.3 Chemical and microbiological analyses

Water samples collected during the experiments were analyzed for the redox relevant species SO_4, Fe, CH_4, and DOC. Fe was analyzed as total dissolved Fe (following filtration over a 0.45 μm filter), speciation calculations with PHREEQC v2.15 (Parkhurst & Appelo, 1999) showed this is >99% Fe(II). Terminal restriction fragment length polymorphism (T-RFLP) and 454 pyrosequencing of PCR-amplified bacterial 16S rRNA genes and Denaturing Gradient Gel Electrophoresis (DGGE) of PCR fragments of archaeal 16S rRNA genes followed by sequencing of excised bands were used to detect changes in the microbial community. Details on analytical procedures are provided in the appendix of this chapter.

5.2.4 Deriving kinetic and thermodynamic parameters

To determine the temperature dependence of sulfate-reduction coupled to sedimentary organic matter (SOM) mineralization, we first derived the sulfate reduction rates from the observation data using a Monod kinetics rate equation (Benner *et al.* 2002; Appelo & Postma, 2005; Pallud & Van Cappellen, 2006; Pallud *et al.* 2007; Tarpgaard *et al.* 2011; Sharma *et al.* 2012):

$$\mathrm{SRR}(T) = \mathrm{SRR}_{\max}(T)\frac{m_{\mathrm{SO_4}}}{k_{\mathrm{SO_4}} + m_{\mathrm{SO_4}}} \tag{5.1}$$

where SRR(T) is the sulfate-reduction rate (M s^{-1}) at temperature $T(K)$, $\mathrm{SRR}_{max}(T)$ is the maximum rate constant (M s^{-1}) at temperature $T(K)$ not restricted by lacking sulfate, $m_{\mathrm{SO_4}}$ is the sulfate concentration

(M), and k_{SO_4} is the half saturation constant [M]. The temperature dependence of SRR_{max} can be expressed using the Arrhenius equation (Overmann, 2013):

$$\ln[SRR_{max}(T)] = \ln(A) + \left(\frac{-E_a}{R}\frac{1}{T}\right) \quad (5.2)$$

where E_a is the activation energy (J mol^{-1}), R is the molecular gas constant (8.314 J K^{-1} mol^{-1}), T is the absolute temperature (K), and A is pre-exponential factor [M s^{-1}]. Note that only the activation energy is required to convert a rate from one temperature to another. The temperature ramping experiment allowed determination of the optimum temperature for sulfate-reduction (T_{opt}, K) and SRR at this temperature ($SRR_{T_{opt}}$, [M s^{-1}]), which can be compared to the SRR at in-situ temperatures ($SRR_{in\text{-}situ}$, [M s^{-1}]). *A* commonly used parameter to express the temperature dependence is Q_{10} which gives the increase in reaction rate for a 10 K increase in temperature (Overmann, 2013):

$$Q_{10} = \exp\left[\frac{E_a 10}{RT(T+10)}\right] \quad (5.3)$$

Equations (5.1) to (5.3) were used to determine $SRR_{T_{opt}}$, $SRR_{in\text{-}situ}$, E_a, and Q_{10} values by minimizing the difference between observed and simulated SO_4 concentrations using the universal model optimizer PEST (Doherty, 2010). The half saturation constant, k_{SO_4}, was based on the transition from a linear to first-order sulfate-reduction rate as observed in the SO_4 versus time plots.

5.3 RESULTS

5.3.1 Increasing residence time (IRT) experiments

SO_4 depletion was not significant in sediment H2 at in-situ temperature (11°C) and below (5°C), however, at elevated temperature (25°C and 60°C) SO_4 decreased with increasing residence time indicating occurrence of sulfate-reduction (Figure 5.1a). For sediment S, SO_4 depletion with increasing residence times was observed over the entire temperature range (Figure 5.1b). The elevated Fe(II) concentration after the longest incubation time in the 5°C effluent and to a lesser extent in the 11 and 25°C effluents of sediment H2 indicated the occurrence of iron-reducing conditions (Figure 5.1e). The Fe(II) concentration remained stable in the 60°C effluent of sediment H2, and for sediment S over the entire temperature range (Figure 5.1f).

Effluents were sampled and analyzed for CH_4 from sediment S, and CH_4 was generated in significant amounts only at 25°C, especially during residence times longer than 15 days (Figure 5.1g). In the 60°C column, no CH_4 was produced. When the observed changes in redox species were expressed on the basis of transferred electrons (by multiplying the change in the concentration of electron acceptor with the number of transferred electrons (Friis *et al.* 2006b): 8 for SO_4, 4 for CH_4 and 1 for Fe(II)), the total electron transfer by sulfate reduction (0.57 me L^{-1}, at 60°C in sediment H2) was much higher than by iron reduction (0.03 me L^{-1}, at 5°C in sediment H2) or CH_4 production (0.008 me L^{-1} at 25°C in sediment S). Because this approach did not consider precipitation of iron-sulfides, the transferred electrons for iron-reduction should be considered as a minimum. In sediment H2, DOC increased at 60°C, whereas in sediment S, an increase in DOC production was observed at both 25°C and 60°C (Figure 5.1c, d).

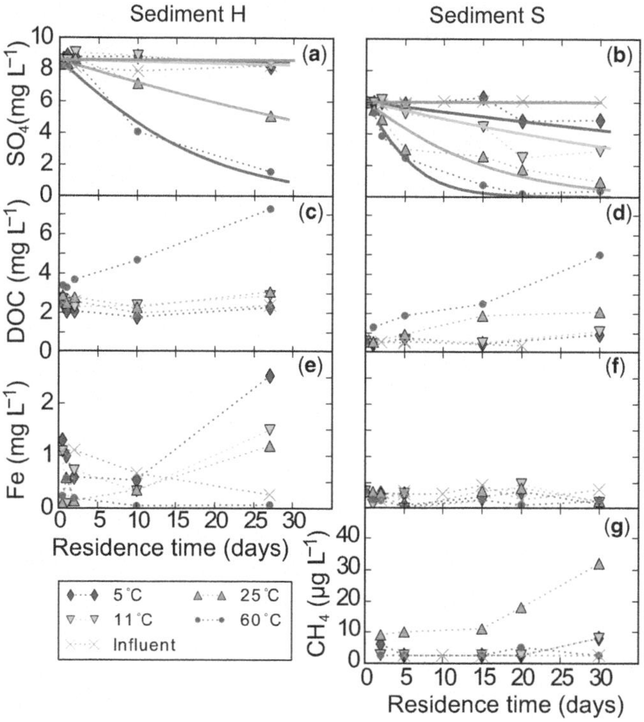

Figure 5.1 Observed concentrations of SO_4, DOC, Fe(II), and CH_4 (symbols and dashed lines) for effluents of sediments H and S as a function of residence time. Observed SO_4 data were also fitted to a Monod kinetic model (solid lines).

5.3.2 Temperature ramping (TR) experiments

The TR experiments showed SO_4 depletion occurring at all temperatures (Figure 5.2a), indicating sulfate-reduction to occur, whereas Fe(II) concentrations remained near the influent concentration (Figure 5.2c), both of which are in line with the results of the IRT experiments. The higher temperature resolution did, however, reveal two distinctive minima in SO_4 concentrations in the effluent, at $T_{opt} = 40$ and at $T_{opt} = 70$°C. Between 45 and 50°C, SO_4 concentrations were very close to the influent concentrations. CH_4 production was greatest between 25°C and 40°C corresponding to the first region of maximum sulfate-reduction (Figure 5.2d). The DOC concentrations were close to the influent concentration up to 15°C (Figure 5.2c). Beyond 20°C, a gradual increase in DOC was observed from 0.5 mg L^{-1} (equal to influent) to 3.6 mg L^{-1} at 80°C. Noteworthy, the general pattern of steadily increasing DOC concentration lacks distinct minima as observed for sulfate.

5.3.3 Microbial community changes

Bacterial 16S rRNA gene-based T-RFLP (Figure S2.2 in Appendix 2) and pyrosequencing (Figure 5.3, Figure S2.3 in Appendix 2; 1670 sequences per sample) revealed obvious differences between the bacterial communities in effluents and sediments. At temperatures between 5 and 25°C, the bacterial communities in the effluent strongly resembled the bacterial community in the influent. However, at 60°C a clear change in community structure was observed, especially for communities derived from sand (Figures 5.3, S2.2, S2.3). The bacterial community shifted from *Proteobacteria*-dominated (50–92% of Operational Taxonomic Units (OTUs)) to *Firmicutes* (26–72%); in particular anaerobic, moderate thermophilic, sporeforming *Thermoanaerobacteraceae* and *Peptococcaceae* (both *Clostridia*) and *Bacilliceae* (Figure S2.3, Table S2.2 in Appendix 2) strongly increased in abundance. Furthermore, an OTU strongest related to *Caldilineaceae* (*Chloroflexi*) increased to 14% of OTUs in sand. These families also became enriched in the effluent.

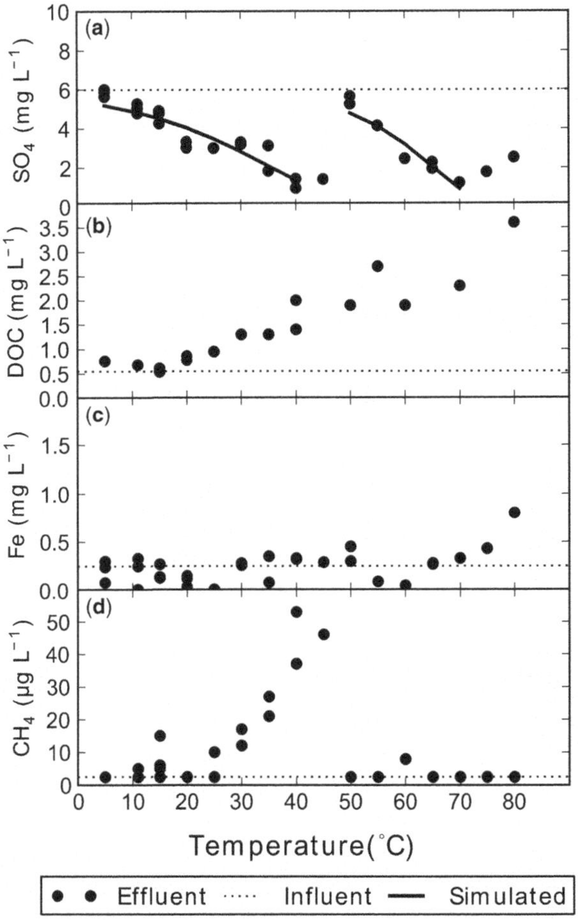

Figure 5.2 Temperature ramping experiment with sediment S. Observed concentrations of SO_4, DOC, Fe(II) and CH_4 (points) for effluent with 5 day residence time at different temperatures. Also shown are the influent concentrations (thin dashed line) and the simulated SO_4 concentrations.

Temperature-related shifts in the occurrence of OTUs that can be linked to specific redox processes were evident. A diversity of sulfate-reducers, belonging to the genera *Desulfovibrio, Desulfatiferula, Desulfobacula, Desulfarculus, Desulfosalsimonas, Desulfobulbus, Desulfurivibrio, Desulfatibacillum* and *Desulfobacca* (all *Proteobacteria*), and *Desulfosporosinus*, and *Desulfurispora* (*Firmicutes*), was present at moderate temperatures, but the sulfate-reducing *Proteobacteria* disappeared at 60°C (Table S2.2 in Appendix 2). At this temperature OTUs closest related to sporeforming, sulfate-reducing and thermophilic *Desulfotomaculum* (*Firmicutes*) appeared, although it should be noted that DNA similarities were very low (87–90%).

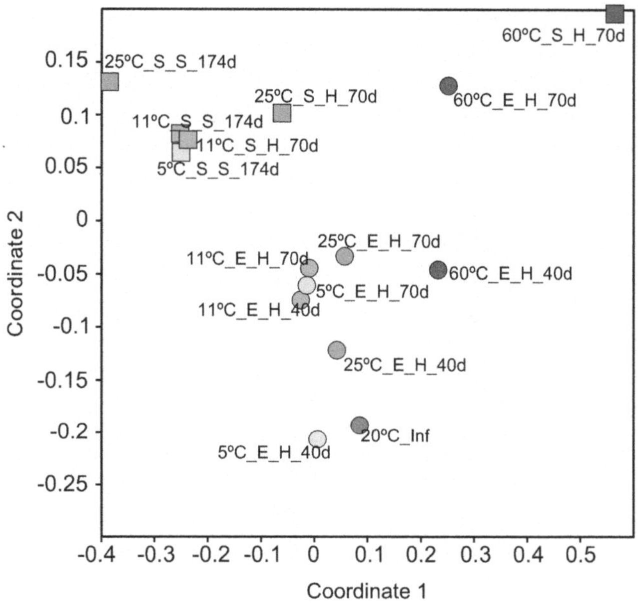

Figure 5.3 Non-metric multidimensional scaling plot of Bray-Curtis similarities among bacterial communities in effluent and sediment in column experiments at various temperatures, using 16S rRNA gene pyrosequencing. Stress was 0.092. Coloring of symbols corresponds to temperature levels used in Figure 5.1, circles represent water samples, squares represent sand samples. Sample codes provide information on temperature (5, 11, 25 or 60°C), medium (Effluent or Sand), sampling location (Helvoirt or Scherpenzeel) and incubation time (in days).

Likewise, iron-reducers belonging to the genera *Geothrix* (*Acidobacteria*), *Shewanella, Desulfuromonas, Geobacter,* and *Abideferax* (all *Proteobacteria*) and *Desulfitobacterium* (*Firmicutes*), completely disappeared at the highest temperature (Table S2.2 in Appendix 2). At 25°C, in particular *Desulfuromonas* and *Geobacte*r were present at lower abundance in sediment, than at 5 and 11°C, in line with the observed lower iron-reduction inferred from hydrochemical data.

Furthermore, temperature had a large impact on *Archaea* communities. While nearly all (9 out of 12) samples of the experiments held at 5 to 25°C revealed an *Archaea*-specific PCR product, only one of the 4 samples from the experiment at 60°C was positive. DGGE analysis revealed a single band, which was 97% similar to the ammonium-oxidizing Thaumarchaeote *Nitrosopumilis* (Figure S2.4 in Appendix 2). In contrast, sequencing of several bands from DGGE profiles corresponding to samples taken from column experiments conducted at lower temperatures revealed methanogenic Euryarchaea, in line with the observed CH_4 production.

5.3.4 Kinetics and thermodynamics of sulfate-reduction

The Arrhenius equation can only be applied over a limited temperature range, depending on the type of enzymes involved and their denaturation temperature, which was not a priori known (Overmann, 2013).

We used the results of the TR experiment (Figure 5.2) and the 454-pyrosequencing data (Figure 5.3) to constrain two temperature ranges (defined here as a mesophilic range of 5–50°C and a thermophilic range of 50–70°C) for which kinetic and thermodynamic parameters could be derived. In Figures 5.1 and 5.2, the simulated SO_4 concentrations are shown on basis of the optimized parameters for the Monod and Arrhenius equations (Equations 5.1 and 5.2), and a visually derived half saturation concentration of 4×10^{-5} M.

In-situ sulfate-reduction rates (SRR) of 4.8×10^{-12} to 4.3×10^{-11} M s^{-1} at 11°C, and E_a values of 115 ± 268 and 54 ± 26 kJ mol^{-1} were calculated for the IRT experiments on sediments H2 and S, respectively (Table 5.1). The mesophilic E_a for sediment H2 had a very large standard error which is due to the low SRR at 5 and 11°C, causing the E_a to be poorly constrained. The TR experiment on sediment S yielded maximum SRR_{opt} values of $4.8 \pm 6.1 \times 10^{-10}$ and $5.3 \pm 11 \times 10^{-10}$ M s^{-1} at T_{opt} of 40°C and 70°C, respectively. Derived E_a-values for the low and high temperature ranges of sediment S were 41 ± 19 and 80 ± 44 kJ mol^{-1}, respectively.

Table 5.1 Kinetic and thermodynamic sulfate-reduction parameters for the monod kinetic model. shown are derived best fit values and the standard errors (±).

Sediment / experiment	T_{opt} (°C)	T_{max} (°C)	$SRR_{in\text{-}situ}$ (M s^{-1})	$SRR_{T_{opt}}$ (M s^{-1})	E_a (kJ mol^{-1})	Q_{10}
A. IRT experiments						
Sediment H2 (5–25°C)	n.a.[a]	n.a.	4.8×10^{-12}	n.a.	115 ± 267	5.3
Sediment S (5–25°C)	n.a.	n.a.	4.3×10^{-11}	n.a.	54 ± 26	2.2
B. TR experiments						
Sediment S (5–40°C)	40	47	4.3×10^{-11}	$4.8 \pm 6.1 \times 10^{-10}$	41 ± 19	1.9
Sediment S (50–70°C)	70	>80	n.a.	$5.3 \pm 11 \times 10^{-10}$	80 ± 44	3.2

Notes: [a]T_{opt}, T_{max} and $SRR_{T_{opt}}$ could only be determined in the temperature ramping experiment: n.a., not available.

5.4 DISCUSSION

5.4.1 Impact of temperature on prevailing redox reactions

Over the entire temperature range in sediment S and at elevated temperature (25 and 60°C) in sediment H2, hydrochemical data, supported by microbial data, revealed sulfate-reduction as the dominant redox reaction, whereas iron-reducing conditions only occurred at in-situ temperature (11°C) and below (5°C) in sediment H2. The prevalence of sulfate-reduction in sediment S and the shift in prevailing redox processes from iron-reduction to sulfate-reduction with increasing temperature observed in sediment H2, can be placed in a thermodynamic context by calculating the H_2 threshold required for a given redox reaction to occur (Hoehler *et al.* 1998; Jakobsen *et al.* 1998; Hansen *et al.* 2001).

Microorganisms that gain energy from redox reactions with a low H_2 threshold have a competitive advantage because they are able to oxidize H_2 produced by fermenting organisms at H_2 levels sufficiently low to impede the respiration of microorganisms that depend on redox processes having higher H_2 thresholds. We calculated the temperature-dependent H_2 thresholds for specific redox-processes from the initial aqueous concentrations, the standard Gibbs free energy, and the minimum free energy required for biochemical reactions to proceed (−7 kJ mol^{-1} based on Jakobsen *et al.* (1998) and van Breukelen *et al.* (2003)) (Figure 5.4, details on the calculation are presented in the SI). The relationship between temperature and redox-process specific H_2 thresholds shows that reduction of Fe(III) from lepidocrocite (a relatively reactive iron-oxide) has the lowest H_2 threshold at a temperature below 25°C, while at higher temperature sulfate-reduction proceeds with the lowest H_2 threshold. This agrees with the observed transition from iron- to sulfate-reducing conditions with elevated temperature in sediment H2 and suggests that both this transition, and the disappearance of the iron-reducing community at 60°C, are induced by thermodynamic constraints.

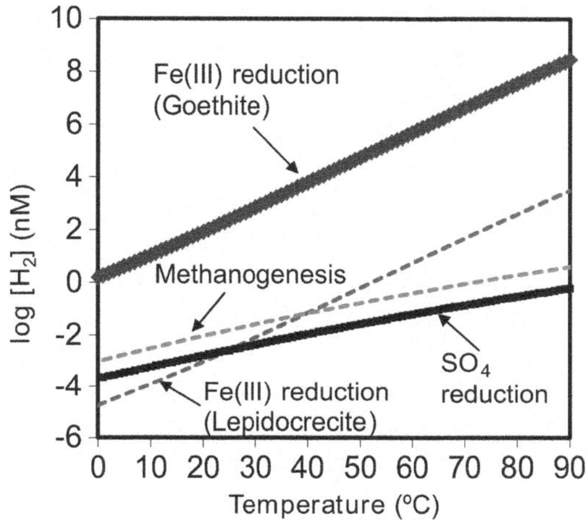

Figure 5.4 Hydrogen gas thresholds for different terminal electron accepting processes as function of temperature using pH = 7.0, Fe(II) = 10^{-4} M, $SO_4^{2-} = 10^{-4}$ M, $HS^- = 10^{-12}$ M, $CH_4 = 10^{-11}$ M, $HCO_3^- = 10^{-3}$ M at 11°C (speciation at other temperatures was calculated using the geochemical speciation model PHREEQC v2.15 (Parkhurst & Appelo, 1999)), and assuming the ΔG_{min} of reaction is –7 kJ mol^{-1}.

Although iron-oxide content was similar for both sediments, dissolved Fe(II) was considerably higher in groundwater of site H (6.5 mg L^{-1} and 0.3 mg L^{-1} at sites H and S, respectively), potentially indicating more favorable conditions for iron-reduction at site H. Possibly, iron-oxides were more reactive at site H (e.g., presence of lepidocrocite) promoting iron-reduction there. Figure 5.4 shows that iron-reduction with less reactive goethite is unlikely to occur at ambient aqueous conditions and explains why sulfate-reduction is dominant at site S.

At the shown aqueous concentration, methanogenesis is not thermodynamically favorable, but as concentrations of SO_4 decrease with on-going sulfate-reduction the H_2 threshold for sulfate-reduction increases. When it approaches the H_2 threshold for methanogenesis, this redox process becomes possible. This is also visible in Figure 5.1b and 5.1g with the rate of methanogenesis increasing with decreasing SO_4 concentrations.

5.4.2 Thermophilic redox processes and microbial communities

Both sulfate-depletion, sulfate-reduction rates, and molecular data indicate the emergence of a thermophilic fermenting and sulfate-reducing bacterial community at temperatures >45°C. This shows that part of the microbial functionality was maintained despite great temperature changes. In contrast, hydrochemical data, supported by microbial data indicated that iron-reduction and methanogenesis were insignificant in the thermophilic range. A number of studies in Arctic marine sediments showed a similar emergence of a thermophilic fermenting and sulfate-reducing bacterial community in laboratory experiments mimicking the mesophilic microbial metabolism (McBee & McBee, 1956; Isaksen *et al.* 1994; Bae *et al.* 2005; Hubert *et al.* 2009; Hubert *et al.* 2010). Hubert *et al.* (2010) used SRRs and microbial data to study the temperature impact on organic matter mineralization in marine sediments and found two distinct SRR peaks at 32 and 56°C. These T_{opt} values were both around 10°C lower than found here, which corresponds to the difference in in-situ temperatures between ours and their site. On basis of the analysis of cloned 16S ribosomal RNA genes, Hubert *et al.* (2009) further showed that at higher temperatures (50°C), a thermophilic community developed with *Caminicella* and *Caloranaerobactor-Clostridiisalibacter-Thermohalobacter* lineages (*Firmicutes* phylum) capable of producing volatile fatty acids. This production stimulated the germination of spores and growth of *Desulfotomaculum* species; sulfate-reducing, spore-forming bacteria which were naturally present in these sediments at very low concentrations (Hubert *et al.* 2009).

The emergence of several OTU closest related to spore-forming, thermophilic members of the *Firmicutes* phylum was also evident in our sediments despite the very different origin of sediments used by Hubert *et al.* (2009) and ours. The origin of thermophilic spore-formers found by Hubert *et al.* (2009) was speculated to be petroleum reservoirs or mid-oceanic spreading centers from which spores can migrate into superficial sediments. (Hubert *et al.* 2009) Our sediments were taken from an active geological graben, bound by faults which penetrate to several kilometers depth (van Balen *et al.* 2005). These faults are characterized by a high anisotropy with high permeability along the fault plain and reduced perpendicular permeability (Bense *et al.* 2003; Bense & Van Balen, 2004). We speculate that, potentially, these faults provide preferential flow paths for thermophilic spores to migrate to shallower geological layers.

5.4.3 Accumulation of dissolved organic carbon (DOC) and organic carbon turnover

Mineralization of sedimentary organic matter (SOM) comprises a number of steps (Muyzer & Stams, 2008) including hydrolysis, fermentation and terminal oxidation, for example, by sulfate-reduction, while the intermediate reaction products contributing to DOC are generally kept at low concentrations at the in-situ temperature (Robador *et al.* 2011). This is consistent with our observation that in sediment S, sulfate-reduction was occurring at 5°C, 11°C, and 15°C, but no increase in DOC was observed (Figure 5.2). Only if the rate of one of the first two steps (hydrolysis and fermentation) exceeds the rate of the terminal oxidation step, an increase in DOC is expected. Such a decoupling between production and consumption rates and associated accumulation of DOC, was observed in marine sediments when incubated above their in-situ temperature (Weston & Joye, 2005; Robador *et al.* 2011). This decoupling has been postulated to be due to either an incapability of the growth of a fermentative population, or DOC being less bioavailable at higher temperature (Robador *et al.* 2011). The latter explanation is more likely in our case as we observed the emergence of a thermophilic fermentative bacterial community.

The changing reactivity with temperature of mobilized DOC was also observed by Xu and Saiers (2010) in unsaturated soil column experiments and by Brons (1992) in saturated column experiment. Xu and Saiers (2010) explained this by the dependence of DOC mobility on both the temperature and molecular weight. Relatively large (recalcitrant) organic molecules remain sorbed at lower temperature implying that the accumulation of DOC with rising temperature, is in fact a physical or a-biotic process (Xu & Saiers, 2010). The fairly continuous DOC increase over the entire temperature range (Figure 5.2), without the clear transition of mesophilic to thermophilic activity as observed in the sulfate depletion, substantiates that the DOC formation is primarily an abiotic process producing relatively recalcitrant DOC, not readily consumed by the fermenting community. This results in the accumulation of DOC with both temperature and residence time.

5.4.4 Sulfate-reduction kinetics and thermodynamics

A comparison between in-situ SRRs reported in the literature for aquifers (Figure 5.5a and Table S2.3 in Appendix 2, with data by Jakobsen & Postma, 1994, 1999, and Benner *et al.* 2002)) and our results shows that the SRR for sediment H2 falls within the 50% central region of SRR-values reported in literature, while that of sediment S is just above this. It is also evident from Figure 5.5a that SRRs observed in aquifers span a large range with values ranging three orders of magnitude. SRR in aquifer sediments calculated by Jakobsen and Postma (1994) showed a general tendency of the oldest sediments yielding the lowest SRRs. Our sediments are of Pleistocene age, and SSRs specific for this age are at the top end of the reported range. The relatively high SRRs especially for sediment S could be attributed to the fact that the SRRs reported by Jakobsen and Postma (1994) were based on field estimates, which generally result lower values than laboratory estimates (Hansen *et al.* 2000). The SRRs in fresh aquifer sediments are generally lower than those observed in both saline marine (Isaksen *et al.* 1994; Arnosti *et al.* 1998; Marvin-DiPasquale & Capone, 1998; Roychoudhury *et al.* 1998; Roychoudhury, 2004; Pallud & Van Cappellen, 2006; Pallud *et al.* 2007; Robador *et al.* 2009; Sawicka *et al.* 2012) and geothermal sediments (Jørgensen *et al.* 1990; Elsgaard *et al.* 1994; Kallmeyer & Boetius, 2004) (Figure 5.5a), which may due to a combination of sediment age (Middelburg, 1989; Jakobsen & Postma, 1994), the extent of oxygen exposure during deposition (Hartog *et al.* 2004), and the lower abundance of recalcitrant biomacromolecules (lignin) in marine sediments (Hartog *et al.* 2004).

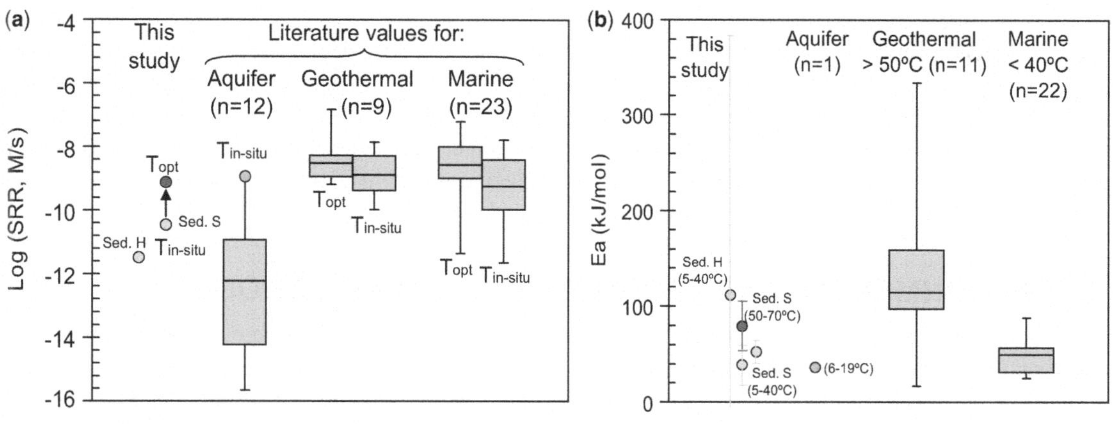

Figure 5.5 Panel a shows sulfate-reduction rates (SRR) for our experiments at $T_{in\text{-}situ}$ and T_{opt} (light gray and dark gray dots, respectively) compared with box plots of SRR for in-situ and optimum temperatures derived from literature for aquifer (Jakobsen & Postma, 1994, 1999; Benner *et al.* 2002) (in-situ only), marine (Arnosti *et al.* 1998; Marvin-DiPasquale & Capone, 1998; Roychoudhury *et al.* 1998; Roychoudhury, 2004; Pallud & Van Cappellen, 2006; Pallud *et al.* 2007; Robador *et al.* 2009; Sawicka *et al.* 2012), and geothermal sediments (Jørgensen *et al.* 1990; Elsgaard *et al.* 1994; Kallmeyer & Boetius, 2004). Panel b shows the activation energy (E_a) of sulfate-reduction derived from our experiments (dots), a reactive permeable barrier in an aquifer (Benner *et al.* 2002), and box plots for values for geothermal (Jørgensen *et al.* 1990; Elsgaard *et al.* 1994; Kallmeyer & Boetius, 2004), and marine sediments. (Isaksen *et al.* 1994; Marvin-DiPasquale & Capone, 1998; Hubert *et al.* 2009; Sawicka *et al.* 2012) E_a values for a number of these studies (Jørgensen *et al.* 1990; Isaksen *et al.* 1994; Kallmeyer & Boetius, 2004; Hubert *et al.* 2009) were derived from SRR at different temperatures. The SRR and E_a values found by Benner *et al.* (2002) (discussed in text) are indicated by the mid grey point. Tabulated data from literature is presented in Table S2.3 in Appendix 2.

Modeling temperature-dependent sulfate-reduction in aquifers using the Arrhenius equation requires an E_a value. Such value has to our knowledge only been reported by Benner *et al.* (2002) for a reactive barrier comprising reactive organic matter from municipal compost aiming to remediate a nickel contamination placed in an aquifer ($E_a = 40$ kJ mol^{-1}; Figure 5.5b). This value is lower than the mesophilic range found for sediment S (54 ± 26 and 41 ± 19 kJ mol^{-1} for the IRT and TR experiments, respectively) and for sediment H2 (115 ± 268 kJ mol^{-1}). Comparing the SRR and E_a values shown in Figure 5.5 with our results, there is a tendency that sulfate-reduction coupled to relatively recalcitrant SOM oxidation (our experiments) proceeded at a low SRR and was relatively sensitive to temperature change (high E_a and Q_{10} values) in comparison to younger and more reactive C present in the reactive barrier described by Benner *et al.* (2002) and the marine studies shown in Figure 5.5. This agrees with the conceptual model for temperature dependence of soil carbon mineralization by Davidson and Janssens (2006), in which the mineralization rate of recalcitrant soil organic matter is low but highly temperature-dependent.

A number of studies on arctic marine sediments (Marvin-DiPasquale & Capone, 1998; Sawicka *et al.* 2012) report E_a values of sulfate-reduction in the same range as observed for the mesophilic range in sediment S (50% of E_a values between 30 and 60 kJ mol^{-1} in Figure 5.5b). The E_a for sulfate-reduction in geothermal systems shown in Figure 5.5b, however, reveals an E_a range between 100 and 160 kJ mol^{-1} (50% box) suggesting a different thermodynamic behavior of thermophilic compared to mesophilic sulfate-reduction. We observed a similar effect in the TR experiment using sediment S, where the E_a was considerably larger in the thermophilic (80 kJ mol^{-1}) than in the mesophilic range (44 kJ mol^{-1}). A possible explanation for the larger activation energy under thermophilic conditions is a trade-off in biochemical processes between the enzyme stability and activity (D'Amico *et al.* 2003). The catalytic capacity of enzymes depends on their molecular flexibility which determines their capacity to bind and catalyze substrates (Danson *et al.* 1996). Enzymes produced by thermophilic sulfate-reducing organisms require a higher thermostability, making them less flexible and less efficient, potentially resulting in higher E_a values for thermophilic enzymes. This phenomenon has not yet been reported for microbial processes in aquifers, but has been reported for a pure culture study using mesophilic and thermophilic homologues (Lam *et al.* 2011) and in another study on a number of psychrophilic enzymes (Feller, 2003).

5.5 ENVIRONMENTAL AND TECHNOLOGICAL IMPLICATIONS

Overall, our chemical and molecular data, combined with the kinetic and thermodynamic calculations show that the impacts of shallow geothermal energy (or any other temperature perturbation) on key biochemical functions of the aquifer depend on (i) the thermodynamic constraints on H_2 substrate utilization causing under our conditions in sediment H2 (having reactive iron-oxides) a shift from iron-reducers to sulfate reducers, (ii) the temperature tolerance of the in-situ microbial community, which determines the range up to which a temperature increase from in-situ conditions will result in acceleration of biochemical processes, (iii) the presence of spores of thermophilic bacteria, determining whether a thermophilic microbial community will develop following the die-off of the initially present microbial community, and (iv) the in-situ SOM reactivity, where a lower reactivity generally results in higher E_a and Q_{10} and thus a more pronounced effect of temperature increase.

The observed temperature-induced impacts on redox processes are relevant for groundwater quality, drinking water production, and biodegradation of organic pollutants. Firstly, the increasing mineralization rate and increasing DOC concentrations with enhanced temperature can cause discoloration of the groundwater requiring additional treatment when used for drinking water production (Amy *et al.* 1992). This process can be especially problematic because DOC mobilized at higher temperatures is potentially more recalcitrant making it hard to remove in water treatment. Secondly, the aquifer's SOM is an important contributor to contaminant buffering capacity (Griffioen *et al.* 2012), for example, as a sorption medium for organic micro-pollutants (Appelo & Postma, 2005). Thermal removal of SOM from aquifers can, therefore, increase their vulnerability. Thirdly, a temperature-induced shift from iron-reducing to sulfate-reducing conditions may affect the biodegradation potential of organic pollutants in various ways. On the one hand, whereas iron-reduction is a key process governing anaerobic degradation of aromatic hydrocarbons, a transition to more reduced conditions was observed to lower their degradation rate (Cozzarelli *et al.* 2001). On the other hand, such a shift and lowering of sulfate levels may promote reductive dechlorination of chlorinated ethenes (Chambon *et al.* 2013).

The identified impacts are also relevant for the design and operation of the shallow geothermal system itself. Sulfate-reduction can cause anaerobic corrosion to, and subsequent failure of, ferrous metals, and to a lesser degree to stainless steel (Brons & Zehnder, 1990; Lerm *et al.* 2013) reducing the life time of the system. Spores of thermophilic sulfate-reducers might be omnipresent in the subsurface, and as such this issue should be considered during design and operation of shallow geothermal systems (Hubert *et al.* 2009). Furthermore, the mobilized organic carbon may provide substrate for microbial communities present on the geothermal well screen, resulting in slime formation and clogging of wells or in line filters (Hijnen & Van der Kooij, 1992; Vetter *et al.* 2012). The latter aspect depends largely on the reactivity of mobilized DOC which, as discussed, is likely to decrease with increasing temperature.

Chapter 6

Reactive transport modeling of thermal column experiments to investigate the impacts of aquifer thermal energy storage on groundwater quality

ABSTRACT

Aquifer Thermal Energy Storage (ATES) systems are increasingly being used to acclimatize buildings and are often constructed in aquifers used for drinking water supply. This raises the question of potential groundwater quality impact. Here, we use laboratory column experiments to develop and calibrate a reactive transport model (PHREEQC) simulating the thermally induced (5–60°C) water quality changes in anoxic sandy sediments. Temperature-dependent surface complexation, cation-exchange, and kinetic dissolution of K-feldspar were included in the model. Optimization results combined with an extensive literature survey showed surface complexation of (oxy)anions (As, B, and PO_4) is consistently exothermic, whereas surface complexation of cations (Ca and Mg) and cationic heavy metals (Cd, Pb, and Zn) is endothermic. The calibrated model was applied to simulate arsenic mobility in an ATES system using a simple yet powerful mirrored axi-symmetrical grid. Results showed that ATES mobilizes arsenic towards the fringe of the warm water bubble and the center of the cold water bubble. This transient re-distribution of arsenic causes its aqueous concentrations in the cold and warm groundwater bubbles to become similar through multiple heating cycles, with a final concentration depending on the average injection temperature of the warm and cold ATES wells.

6.1 INTRODUCTION

Aquifer Thermal Energy Storage (ATES) systems use the underground to store or harvest renewable energy for heating or cooling purposes (Meyer & Todd, 1973). Utilization of ATES is rapidly growing in western European countries, and especially in the Netherlands where the number of licensed ATES systems increased from 29 systems in 1995 to around 1800 systems in 2012 (Bonte, 2013 (Chapter 1)). ATES systems typically comprise two wells (called a well doublet), which serve both as extraction and injection well, depending on the season. In summer, groundwater is extracted from the 'cold' well and used to cool buildings. This raises the temperature of the groundwater which is subsequently

Reprinted with permission from Reactive Transport Modeling of Thermal Column Experiments to Investigate the Impacts of Aquifer Thermal Energy Storage on Groundwater Quality, Matthijs Bonte, Pieter Stuyfzand, and Boris Maurijn Van Breukelen, Environ. Sci. Technol., Just Accepted Manuscript, DOI: 10.1021/es502477m Publication Date (Web): September 24, 2014,

injected in the 'warm' well. In winter, the flow direction reverses and groundwater is extracted from the warm well and used to heat the building. This lowers the temperature of the groundwater before it is re-injected into the cold well. The maximum injection temperature in countries like the Netherlands, Germany, and Austria is legally limited to 20–25°C (Hähnlein *et al.* 2010) but most systems work up to 15°C. Systems with higher operational temperatures (up to 110°C) are less frequently reported, (Perlinger *et al.* 1987; Holm *et al.* 1987) but interest in these systems is growing as a technology of storing waste heat for later re-use. ATES systems are often realized in the same aquifers used for drinking water production, which raises questions about the effects of ATES on water quality (Bonte *et al.* 2011; Ferguson, 2009).

Published research showed ATES can influence groundwater quality by mixing of different water quality types, (Bonte *et al.* 2011b, Chapter 3) temperature changes, (Perlinger *et al.* 1987; Holm *et al.* 1987) or intrusion of oxygen or degassing (Palmer & Cherry, 1984; Jenne *et al.* 1992). Most of these studies focused on operational aspects, such as scaling in relation to mineral precipitation (Perlinger *et al.* 1987; Holm *et al.* 1987; Palmer & Cherry, 1984; Griffioen & Appelo, 1993). A number of laboratory studies showed that the microbial community structure and groundwater fauna was not significantly impacted by temperature increases up to 30°C (Bonte *et al.* 2013b, Chapter 5) Brielmann *et al.* 2009, 2011), which was confirmed in data from a field ATES site (Brielmann *et al.* 2009). However, the rates of kinetically restricted biochemical processes do show a clear temperature dependency, as observed in laboratory experiments for sulfate-reduction (Bonte *et al.* 2013b, Chapter 5) Jesußek *et al.* 2012) and oxidation of ferrous iron (Vollrath *et al.* 2013), and in the field for pyrite oxidation (Prommer & Stuyfzand, 2005).

In Chapter 4, we employed column experiments to investigate the possible effects of thermal changes in anoxic unconsolidated sandy sediments and showed that particularly arsenic and to a lesser degree boron, silica, potassium, phosphorous, and dissolved organic carbon (DOC) became increasingly mobile upon a temperature increase (Bonte *et al.* 2013a, Chapter 4) Also, sulfate-reduction showed a strong temperature-dependence (Bonte *et al.* 2013b, Chapter 5). The aim of the current work is to simulate the experimental results of the column tests (Bonte *et al.* 2013) with a hydrogeochemical reactive transport model using the PHREEQC (Parkhurst & Appelo, 1999) code to gain a quantitative understanding of the temperature dependence of silicate weathering, cation exchange, and surface complexation reactions. We extended the Dzombak and Morel (Dzombak & Morel, 1990) (D&M) database, available in PHREEQC for modeling surface complexation, to include reaction enthalpies for a selection of relevant solutes, which to our knowledge has not been attempted or published before. The calibrated model was subsequently applied to simulate arsenic mobility in a representative ATES system at three injection temperatures using a novel computationally efficient modeling concept composed of two mirrored axi-symmetrical flow tubes. This setup allows for gaining a fast and numerically efficient understanding of the prevailing hydrochemical processes in an ATES system without the need of full 3-D hydrogeochemical transport modeling.

6.2 EXPERIMENTAL METHODS

The experimental methods and data, and conceptual interpretation, used to construct the reactive transport model is described in detail in Chapter 4. A comprehensive summary is provided here. Sediments and water used in the column experiments were collected at Scherpenzeel (Netherlands) from a depth of 34–36 m-SL, which is sediment sample S as used in Chapter 4. The sediment was collected from an unconsolidated sandy aquifer (Sterksel formation, Early to Middle Pleistocene, Westerhoff, 2003). The sample consisted mainly of quartz sand (>90%), with minor fractions of K-feldspar, clay minerals, organic matter, carbonates, pyrite, and reactive iron (Table 6.1 & Table S3.1 in Appendix 3). Following sample collection and homogenization in a glovebox under N_2 atmosphere, four identical sediment columns (with length = 440 mm and diameter = 66 mm) were placed in the experimental setup (Figure S3.1) at temperatures of 5°C (representing cold storage), 11°C (ambient temperature), 25°C (maximum allowed regular ATES), and 60°C (high temperature ATES). Influent (Table 6.1) was derived from groundwater collected from the same borehole from which the sediment cores were taken.

The columns were flushed and the effluents were frequently sampled. The residence time in the columns was 1 day and a thermodynamic equilibrium was assumed for cation-exchange and surface complexation. Afterwards, a tracer test was conducted using NaCl to determine the dispersivity (α) in the cores. This yielded $\alpha = 9$ cm for the 11°C column, $\alpha = 10$ cm for the 5°C and 25°C columns, and $\alpha = 24$ cm for the 60°C column (Figure S3.2 and Table S3.2 in Appendix 3).

6.3 MODELING FRAMEWORK

6.3.1 Modeling framework, boundary, and initial conditions

We used PHREEQC version 2.18 (Parkhurst & Appelo, 1999) to simulate temperature-dependent surface complexation, cation exchange, carbonate dissolution and kinetic K-feldspar weathering. Thermodynamic data were used from the Minteq.v4 (Parkhurst & Appelo, 1999) database, amended with the cation exchange data and silicate mineral dissolution rate equations from the PHREEQC (Parkhurst & Appelo, 1999) database. We applied two model setups: (a) a flow tube model for model calibration to the column experiments (hereafter called calibration model), and (b) a double axi-symmetrical flow tube model (here after called 'DAFT' model) to simulate an ATES system consisting of a connected well doublet including the hydrochemical processes the calibration model (Figure 6.1).

The space and time discretizations of the calibration model for the column experiments were $\Delta x = 8.8$ cm (5 cells) and $\Delta t = 24\ \text{h} \times 60\ \text{min} \times 60\text{s}/5\ \text{cells} = 17{,}280$ s, respectively. The cell size (Δx) was based on the selection of a minimum number of flow cells such that $\Delta x \leq \alpha$, required for a sufficiently low Peclet number to minimize numerical dispersion and oscillations (Parkhurst & Appelo, 1999), and limited to the lowest found dispersivity of $\alpha = 9$ cm. The time step Δt followed from the applied residence time in the columns of 1 day.

The DAFT model comprised two connected, mirror imaged axi-symmetrical flow tubes of 50 cells for each ATES well (Figure 6.1). The connecting 51st cell represents the heat exchanger present between the ATES system's cold and hot wells. Radial flow was simulated by decreasing the cell lengths away from the center cell according to Equations 6.1 and 6.2 by Appelo and Postma (Appelo & Postma, 2005):

$$\Delta x_{i=1} = \frac{L}{\sqrt{n}} \tag{6.1}$$

$$\Delta x_i = \Delta x_{i=1}(\sqrt{i} - \sqrt{i-1}) \tag{6.2}$$

where $\Delta x_i{=}_1$ and Δx_i are the cell sizes of the first and ith cell, respectively, n is the total number of cells which is equal to 50 (for each well), and L is the total length of the flow tube, which is equal to 30 m. The calibrated hydrochemical model was applied to an ATES system using the DAFT model grid, simulating temperature and arsenic concentrations over a 20 year period. Arsenic was selected for these simulations because in many regions it is a problematic solute in groundwater due to its high toxicity with a drinking water guideline value of 10 μg L^{-1} set by the World Health Organization (Smedley & Kinniburgh, 2002).

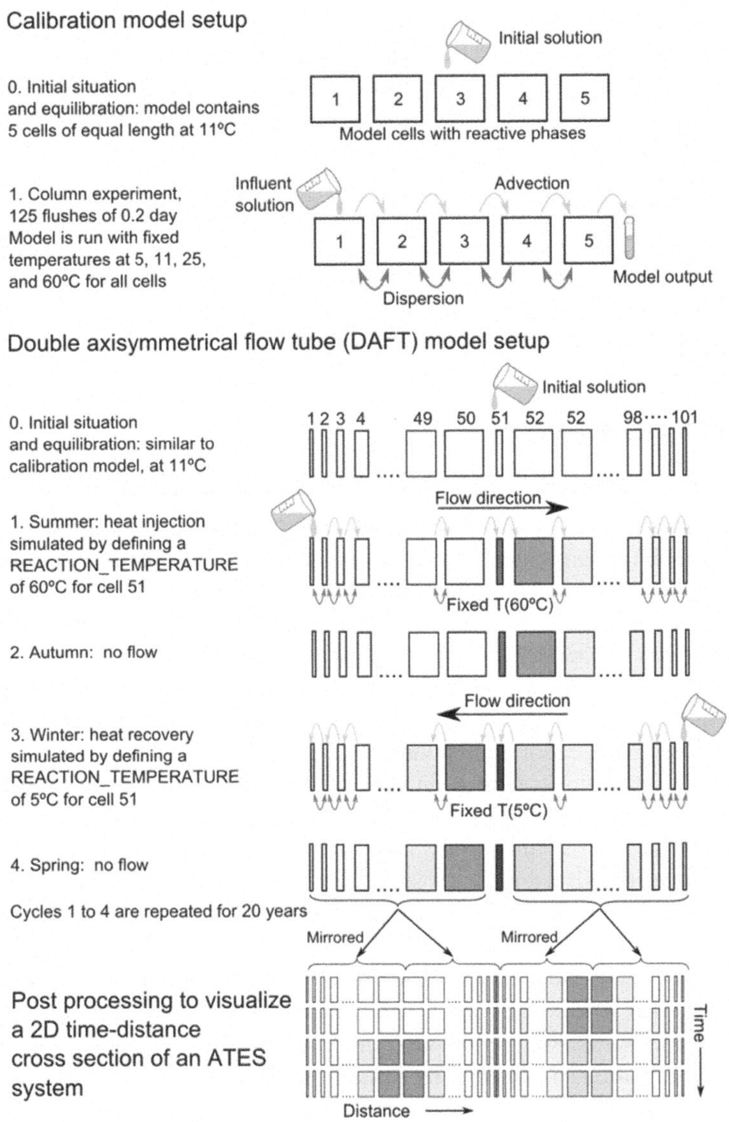

Figure 6.1 Schematic setup of model grids for the calibration model and the DAFT model, and the post processing of DAFT model output.

The initial conditions in both models for the reactive surfaces (cation-exchanger and surfaces) were set by first applying equilibrium between the reactive surfaces and an initial aqueous solution at 11°C (in-situ groundwater temperature). The initial pore water solution for the calibration model was based on the average effluent concentration of the first three 'pore flushes' (i.e., replacing the water volume in the columns three times) of the column at 11°C (in-situ aquifer temperature). Table 6.1 presents an overview of the initial pore water and influent solute concentrations used in the model. For the DAFT model we used the influent concentrations shown in Table 6.1.

Table 6.1 Initial pore water, influent and sediment compositions used in the PHREEQC simulations. Values are in mg/L for solutes and mol/L for minerals, except when stated otherwise.

Water	Initial pore water	Influent
pH	7.2	7.2
Alkalinity (meq/L)	1.8	1.8
Dissolved Oxygen	<0.05	<0.05
NO_3	<0.25	<0.25
NH_4	0.09	0.09
SO_4	6	6
P	0.05	0.05
DOC	2.2	0.55
K	1.4	0.8
Si	6.4	6.4
Ca	37	30
Mg	2.8	2.2
Sr	0.09	0.09
Fe(II)[a]	0.3	0.3
Mn	0.06	0.08
As(III)[a]	0.015	0.001
B	0.05	0.002
Sediment (initial only)[b]		
SOM	0.29 mol L^{-1} (0.14 % d.w.)	
K-feldspar	0.56 mol L^{-1} (2.5 % d.w.)	
Calcite	0.16 mol L^{-1} (0.74 % d.w.)	
CEC	47 meq L^{-1} (7.1 meq kg^{-1})	
Fe-oxides	0.30 mol L^{-1} (0.43 % d.w.)	

[a]Speciation was calculated with PHREEQC showing that >99% of the species occurred in the redox state as shown. [b]Converted from % d.w. to mol/L [*i*] by taking: mol/L = % d.w. $\rho_{bulk}/(n\ M_i)$ with bulk density ρ_{bulk} = 2.12 kg L^{-1} and water content n = 0.32, i is solid phase with mol mass M_i. More details on geochemical parameters are included in Table S3.1 in Appendix 3.

6.3.2 Heat boundary conditions and transport

The calibration model was run at set temperatures of 5, 11, 25, and 60°C as applied experimentally. The DAFT model contained one fixed temperature cell: the mirror cell of the flow tube representing the heat exchanger. Simulations using the DAFT model were carried out for three systems: (i) a 5°C/60°C heat storage system, which in the Netherlands is generally only allowed in deep saline aquifers, (ii) a 5°C/25°C heat storage system, which is the maximum legally allowable temperature in the Netherlands, and (iii) a

7°C/15°C heat storage system which is representative for the design temperature levels for the majority of ATES systems realized in the Netherlands (IF Technology, 2007). Each system (a) injects warm water for 3 months, (b) has a warm water storage phase of 3 months, (c) recovers the warm water and injects cold water for 3 months, and then (d) has a storage phase of 3 months after which the cycle repeats itself.

Heat transport was simulated as a separate species with a thermal retardation factor, R_T, calculated with:

$$R_T = 1 + \frac{(1-n)\rho_s\kappa_s}{n\rho_w\kappa_w} \tag{6.3}$$

where n is the porosity, ρ and κ are the density and specific heat with subscripts s and w denoting the solid matrix and water, respectively. Values of $n = 0.32$, $r_s = 2.65$ kg dm^{-3}, $k_s = 0.830$ kJ K^{-1} kg^{-1}, $r_w = 1.0$ kg dm^{-3}, $k_w = 4.18$ kJ K^{-1} kg^{-1} (based on Engineering toolbox, 2013) yield $R_t = 2.12$.

6.3.3 Temperature correction for mineral equilibria and reaction rates

PHREEQC automatically corrects the equilibrium constants (log K) of mineral interactions and sorption for temperature changes according to the Van't Hoff equation (Appelo & Postma, 2005):

$$\frac{\log K_{T_1}}{\log K_{T_2}} = \frac{-\Delta H_r}{2.303R}\left(\frac{1}{T_1} - \frac{1}{T_2}\right) \tag{6.4}$$

where K_{T1} and K_{T2} are the equilibrium constants at temperatures T_1 and T_2 (K), respectively, and R is the molecular gas constant (8.314 J K^{-1} mol^{-1}). One exception is the temperature dependence of the log K for calcite (discussed further below) which is simulated with a polynomial relationship present in the Minteq.v4 database. Temperature dependence of kinetic reactions was modeled using the Arrhenius equation (Sawicka *et al.* 2012). Converting reaction rates from one temperature T_1 to the next T_2 using the Arrhenius equation can be done with Eq. 6.4 where ΔH_r is replaced with the activation energy E_a (kJ mol^{-1}) and log K is replaced with the log of the reaction rate (log r).

6.3.4 Surface complexation modeling

Surface complexation modeling (SCM) to reactive iron oxide surfaces was performed with PHREEQC using the electrostatic diffuse double layer model of D&M (Table 6.2) (Dzombak & Morel, 1990). Although this model was initially developed for ferrihydrite, the model concept has also been applied to goethite (Appelo *et al.* 2002; Dixit & Hering, 2003) and natural sediments (Stollenwerk *et al.* 2007; Postma *et al.* 2007). Consistent with the D&M database, sorption of anions and oxyanions only occurred on weak sites. We expanded the D&M database with literature values for sorption of silica (Hansen *et al.* 1994; Swedlund & Webster, 1999), carbonate (Appelo *et al.* 2002), and DOC (Evanko & Dzombak, 1998), as these species compete with other species, like arsenic and boron, for sorption sites (Appelo *et al.* 2002; Bauer & Blodau, 2006; Jessen *et al.* 2012). Because DOC comprises a wide variety of organic compounds of different weights (Hartog *et al.* 2004), two DOC weight fractions (DOCA and DOCB with weights of 500 and 1000 g mol^{-1}) were used. Because sorption constants derived for synthetic laboratory grade oxides can differ considerably from those in natural sediments, we optimized sorption constants for silica, arsenic, carbonate, boron, and phosphate.

The original D&M model does not provide ΔH_r values required to include temperature dependence of sorption. However, enthalpy values for surface protonation of oxides are well documented and we used the values derived for goethite from a large set of data and a theoretical model by Sverjensky and Sahai (1998). For the other sorbing species, data in literature is generally less consistent, and we used model optimization to derive sorption enthalpies applying enthalpy values derived from literature (Bonte *et al.* 2013a, Chapter 4; Kersten & Vlasova, 2009b; De Bussetti *et al.* 1995; Trivedi & Axe, 2000) as initial estimates (Table 6.2).

Table 6.2 Surface complexation parameters included in the PHREEQC model with initial and optimized parameter values (± standard error).

Surface complexation reactions	Log K^a		ΔH_r (kJ mol^{-1})^b	
	Initial, other literature values	optimized	Initial	Optimized
$\equiv FeOH + H^+ \leftrightarrow \equiv FeOH^{2+}$	7.29	n.o.[b]	−47.3[c]	n.o.[b]
$\equiv FeOH \leftrightarrow \equiv FeO^- + H^+$	8.93	n.o.[b]	54.0[c]	n.o.[b]
$\equiv FeOH + H_4SiO_4 \leftrightarrow \equiv FeH_3SiO_4 + H_2O$	3.62, 4.28[d]	4.1 ± 0.3	−43.7[e]	−21 ± 4
$\equiv FeOH + H_3AsO_3 \leftrightarrow \equiv FeH_2AsO_3 + H_2O$	5.41	4.9 ± 0.2	−68[g]	−41 ± 5
$\equiv FeOH + H_3BO_3 \leftrightarrow \equiv FeH_2BO_3 + H_2O$	0.62	3.4 ± 0.4	−13.8[h]	−25 ± 20
$\equiv FeOH + CO_3^{2-} + H^+ \leftrightarrow \equiv FeCO_3^- + H_2O$	12.56[i]	n.o.[b]	−68[j]	−49 ± 10
$\equiv FeOH + CO_3^{2-} + 2H^+ \leftrightarrow \equiv FeHCO_3 + H_2O$	20.62[i]	n.o.[b]	−68[j]	−32 ± 7
$\equiv FeOH + DOCA^- + H^+ \leftrightarrow \equiv FeDOCA + H_2O$	10.3[k]	11.6 ± 0.3	−68[j]	−14 ± 30
$\equiv FeOH + DOCB^- + H^+ \leftrightarrow \equiv FeDOCB + H_2O$	10.3[k]	12.6 ± 0.3	−68[j]	−97 ± 20
$\equiv FeOH + PO_4^{3-} + 2H^+ \leftrightarrow \equiv FeHPO4^- + H_2O$[f]	25.39	25 ± 0.6	−68[j]	−59 ± 6
$\equiv FeOH + Ca^{2+} \leftrightarrow \equiv FeOCa^+ + H^+$	−5.85	n.o.[b]	35[l]	57.0 ± 30
$\equiv FeOH + Mg^{2+} \leftrightarrow \equiv FeOMg^+ + H^+$	−4.6	n.o.[b]	80[l]	1.4 ± 20

Iron oxide surface characteristics
Site density: SD = 2 sites nm^{-2} = 3.32 μmol m^{-2} based on value for goethite by Dixit and Hering (2003)
Reactive surface area A_{Fe} = 54 m^2 g^{-1} based on Dixit and Hering (2003)
Weight iron oxides W = 26.4 g L^{-1}
Sorption site sites N_w = SD A_{Fe} W = 4.7 mmol L^{-1}

[a]All equilibrium constants were obtained from the PHREEQC database based on the D&M model for hydrous ferric oxides (Dzombak & Morel, 1990), unless stated otherwise. [b]n.o. not optimized. [c]Based on Sverjensky & Sahai (1998). [d]Based on (Hansen *et al.* 1994; Swedlund & Webster, 1999). [e]value for goethite by (Kersten & Vlasova, 2009). [f]only the second protonation complex was significant in the model. [g]Value for same sediments from van't Hoff plot of sorption isotherm coefficients (Bonte *et al.* 2013a). [h]Value for hydrous Al-oxide from De Bussetti *et al.* (1995). [i]Value for goethite from Appelo *et al.* (2002). [j]No value available, assumed equal to As. [k]first and second protonation constant for 2,3 DHBA (a NOM analogue organic acid) for sorption on goethite Evanko & Dzombak (1998). [l]Value for goethite from Trivedi & Axe (2000).

6.3.5 Cation-exchange

Cation-exchange of Na, K, Ca, Mg, Fe, and Sr was simulated using exchange half reactions related to the reference half reaction for Na exchange (Table S3.3 in Appendix 3) and based on the Gaines-Thomas convention (Appelo, 1994). Values for the exchange coefficients were taken from the PHREEQC (Parkhurst & Appelo, 1999) database, which also includes enthalpies for most solutes except for Fe and Mn. The latter solutes were, however, of minor importance in this experimental system and enthalpies were, therefore, not specified for Fe and Mn. The exchange coefficient (log K) and enthalpy (ΔH_r) for potassium exchange were optimized during model calibration. Local equilibrium was assumed to be applicable for the applied 24 hour residence time, as based on several earlier studies (Griffioen & Appelo, 1993; Bond & Phillips, 1990; Jardine & Sparks, 1984; Sawhney, 1966; Sparks & Jardine, 1984; DeSutter *et al.* 2006). The cation-exchange capacity (CEC) was estimated with an empirical regression equation based on organic matter and clay content (Table 6.1, Table S3.1 in Appendix 3.1) (van Helvoort *et al.* 2007).

6.3.6 Kinetic dissolution of K-feldspar

The PHREEQC database contains a rate equation describing the kinetic dissolution of K-feldspar (Sverdrup, 1990) which we used as an initial dissolution rate, $r_{0,\mathrm{Kfsp}}$ (M s^{-1}) at the start of the experiment:

$$\begin{aligned} r_{0,\mathrm{Kfsp}} &= \frac{\lambda A}{V} k_{\mathrm{sum}} \\ &= \frac{\lambda A}{V} \sum k_i a_i^{n_i} = \frac{\lambda A}{V} (k_{\mathrm{H^+}} a_{\mathrm{H^+}}^{n_{\mathrm{H^+}}} + k_{\mathrm{H_2O}} + k_{\mathrm{OH^-}} a_{\mathrm{OH^-}}^{n_{\mathrm{OH^-}}} + k_{\mathrm{CO_2}} P_{\mathrm{CO_2}}^{n_{\mathrm{CO_2}}} + k_{\mathrm{DOC}} a_{\mathrm{DOC}}^{n_{\mathrm{DOC}}}) \end{aligned} \tag{6.5}$$

where λ is the unitless roughness coefficient (ratio of true and geometric surface area (White & Brantley, 2003)), A is the mineral surface area (dm^2 kg^{-1}), V is the volume of water in contact with 1 kg of mineral (dm^3 kg^{-1}), and k_i is the partial dissolution rate (M s^{-1}) for the ith species (H^+, OH^-, CO_2, DOC) catalyzing the overall reaction, multiplied with the aqueous activity a_i raised to an exponential factor n_i. We applied $A = 18$ dm^2 kg^{-1} and $V = 0.16$ dm^3 kg^{-1} based on grain size data (details in the Appendix 3, section S3.4). The parameter λ increases with age and was allowed to vary within reasonable limits to calibrate the model. The rate parameters (k, n, and E_a) and the temperature dependence of all individual rates were taken from Sverdrup (Sverdrup, 1990) (Table S3.4 in Appendix 3). We combined Eq. 6.5 with an empirical relation by White and Brantley (White & Brantley, 2003) to describe the dissolution rate as a function of time to account for a decreasing reaction rate due to precipitation of secondary minerals:

$$r_{\mathrm{Kfsp}}(t) = r_{0,\mathrm{Kfsp}} / t^m \tag{6.6}$$

where $r_{\mathrm{Kfsp}}(t)$ is the dissolution rate (M s^{-1}) after t days, and m is the power exponent, for which we adopted the value of 0.7 derived by White and Brantley (Doherty, 2010) from a compilation of a large dataset of mineral weathering rates.

6.3.7 Mineral interactions in thermodynamic equilibrium

The dissolution of K-feldspar causes pore water to become supersaturated with respect to gibbsite. Therefore, equilibrium with this mineral phase was maintained during simulations (with log K and ΔH_r from the PHREEQC database). Increasing alkalinity and Ca, Mg, and Sr concentrations, combined with the presence of an impure Mg-calcite in the sediments (shown by XRD, Table S3.1 in Appendix 3) pointed to the dissolution of a Mg,Sr-calcite which was, therefore, included in thermodynamic equilibrium. The fractions of Ca, Mg, and Sr in this mineral phase were 0.881, 0.116, and 0.003, respectively, based on the stoichiometric increases of these solutes in effluent compared to influent. Finally, although XRD analyses

(Table S3.1 in Appendix 3) revealed the presence of Na-feldpar (Albite), the similar Na concentrations in influent and effluent at all temperatures imply that this mineral did not react, and it was, therefore, not included in the model.

6.3.8 Automatic model optimization, sensitivity and uncertainty analysis

The non-linear parameter estimation program PEST (Doherty, 2010) was used to constrain parameters for which literature values were absent or scarce. PEST minimizes the sum of the squared differences between observed and modeled hydrochemical data using the Gauss-Marquardt-Levenberg method. The observations were weighted to the inverse of the observation value according to the method described by Dai and Samper (2004). PEST (Doherty, 2010) was also used to (i) perform a sensitivity analyses of the the different model parameters in the calibration model, and (ii) to calculate the 95%-ile uncertainty bandwidth of predicted As concentrations in the ATES DAFT predictive model using the methodology by Vecchia and Cooley (1987) (details on method in Appendix 3). It should be noted that this type of uncertainty analysis only provides an estimate of the predictive uncertainty due to parameter non-uniqueness and does not consider uncertainty in model structure (Reichert & Omlin, 1997). It should thus be regarded as a minimum degree of mathematical uncertainty associated with the optimized solution.

6.4 RESULTS AND DISCUSSION

6.4.1 Surface complexation modeling results

The optimized leaching pattern for species involved in surface complexation (As, B, Si, DOC and P; Figure 6.2) is predicted reasonably well with the optimized model. Note that additional results for species involved in K-feldspar dissolution and cation exchange are presented in the Supplementary Information. The simulations show that the distribution of species sorbed to surface sites changed most noticeably at 60°C, with declining fractions of silica, phosphate, DOC, magnesium, arsenic, and boron, and increasing fractions of carbonate, protons, and calcium (Figure S3.3 in Appendix 3).

There are also some parts of the effluent curves which are simulated less well: for both DOC and arsenic at 60°C, the initial peak near 3 pore flushes and the observations beyond 20 pore flushes are simulated too low. For boron the simulated effluent curves agree with the observations in that most boron is leached from the sediment during the very first pore flushes, but the predicted concentrations at 60°C are generally too low, while those at 5–25 °C are predicted too high. A number of reasons may explain these subtle mismatches: the model simulates one type of exchange site for surface complexation, while in reality different types are likely present including different types of iron or manganese (hydr)oxides and silicate minerals. Furthermore, DOC simulated as two weight fractions, more likely represents a continuum of weight fractions and molecular structures which can cause the simulated leaching pattern to deviate from the observed.

The optimized model yielded surface complexation constants which were generally close to the initial values from the PHREEQC database (Table 6.2; a summary of all optimized parameter values with their parameter sensitivities is provided in Table S3.5 in Appendix 3). Notable exception is that of boron, for which the optimized value was larger than the value by D&M (Dzombak & Morel, 1990) but still within the 99% confidence interval reported by D&M (Dzombak & Morel, 1990) (–2.66 to 3.89). The sensitivity analyses (Table S3.5 in Appendix 3) revealed that model calibration is most sensitive to the surface complexation constants (relative sensitivity ranging between 7% and 24%, followed by the enthalpy of the surface complexation reactions (0.01% to 9.6%), followed by the parameters related to K-feldspar dissolution and cation exchange (1.3–1.9%).

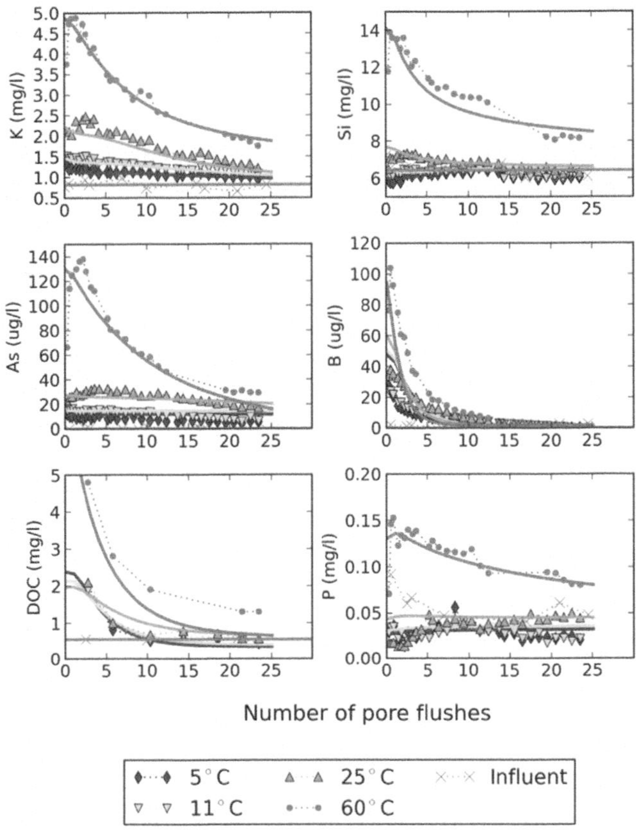

Figure 6.2 Model simulation of column experiments showing a selection of solutes involved in K-feldspar dissolution, cation-exchange, and surface complexation. Lines are simulation results, symbols are observations.

In a comparison of three different surface complexation models (SCMs), Jessen *et al.* (2012) showed that considerable inter-model differences in predicted surface complexation (and thus in surface complexation constants) exist, and that for natural sediments, in particular, which contain various types of reactive surfaces, surface complexation models derived from pure phases have limited accuracy. They therefore recommend calibrating a SCM to data derived specifically for the aquifer under investigation (Jessen *et al.* 2012). This implies that the parameters found here are probably not generically applicable to other aquifers with a different geochemical composition.

6.4.2 Contrasting sorption behavior anions versus cations

The optimized sorption enthalpies for the oxyanions were consistently exothermic (i.e., sorption decreased with increasing temperature) with ΔH_r values ranging from −20 kJ mol^{-1} for silica to −61 kJ mol^{-1} for phosphate, whereas those for the cations calcium and magnesium were endothermic (+5 and +70 kJ mol^{-1}, respectively). Because we only included two cations in our SCM, we investigated the contrasting sorption behavior between cations and anions further by reviewing published enthalpies for cation sorption to various forms of iron and manganese oxides from a large number of different studies (Table S3.6 in Appendix 3). This review revealed that of the 53 reported ΔH_r values for cations, 51 were positive indicating that for most cationic metals sorption increases with temperature so that their mobility reduces with increasing temperature.

The contrasting sorption behavior of cations and anions is also present in the linear free-energy relationships (LFERs) that D&M (Dzombak & Morel, 1990) used in their SCM database. For cations, D&M (Dzombak & Morel, 1990) correlated surface complexation constants with the hydrolysis constants, while (oxy)anions surface complexation constants are correlated with and acidity (protonation) constants. Similar to the sorption enthalpies found here, the first hydrolysis step for the different cations is found by Baes and Mesmer, (1981) to be consistently positive, whereas the enthalpies for the oxyanion protonation reactions (Table S3.7 in Appendix 3, ΔH_r values from the PHREEQC database) are consistently negative.

6.4.3 Application to a virtual aquifer thermal energy storage system

The simulation results for the 60°C system for first 2.5 years show that arsenic is thermally desorbed in the center of the warm bubble and mobilized towards the fringe of this bubble and the center of the cold water bubble (Figure 6.3, Figure 6.4). Correspondingly, sorbed arsenic contents are elevated at the fringe of the warm bubble and particularly in the center of the cold bubble where arsenic accumulates, whereas arsenic showed lowered aqueous concentrations in the center of the warm bubble.

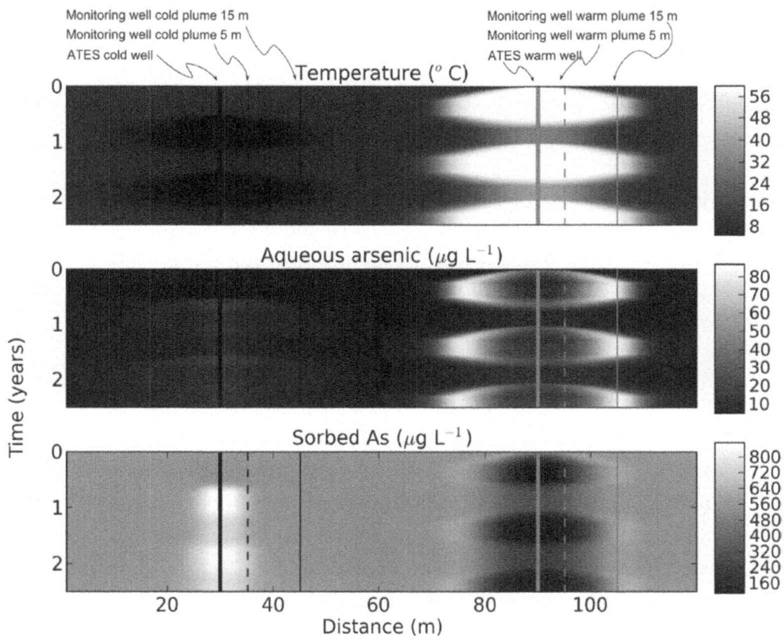

Figure 6.3 Scenario modeling results shown as time-distance plots for temperature, and aqueous and sorbed arsenic for the first 2.5 years of simulation, using the DAFT setup to simulate a 60°C/5°C ATES. Initial groundwater temperature is 11°C, and initial [As] = 5 μg L^{-1}. The warm (60°C) well is indicated with the solid line at X = 90 m, the cold (5°C) well is indicated with the solid line at X = 30 m. Also indicated the location of two monitoring wells at X = 35 m and X = 45 in the cold bubble, and at X = 95 m and X = 105 m in the warm bubble for which the concentrations are shown in Figure 6.5.

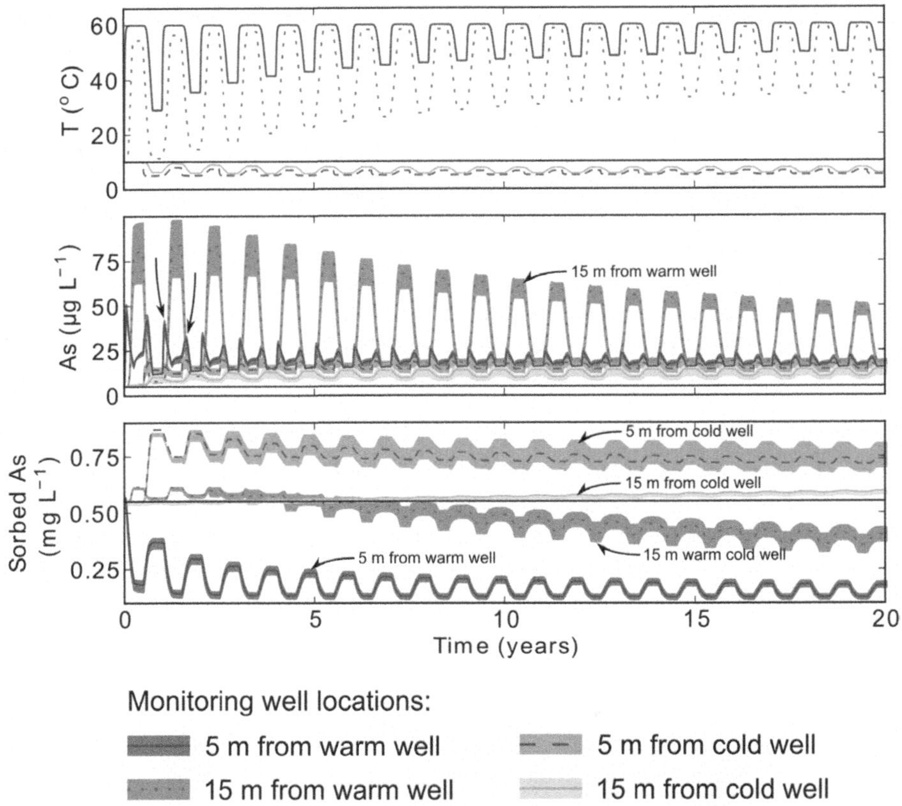

Figure 6.4 Simulated aqueous and sorbed arsenic concentrations in virtual monitoring wells. Predictive uncertainty ranges calculated with PEST (Doherty, 2010) are shown as light shadings. Injection temperature in the cold well was 5°C and 60°C in the warm well. Initial groundwater temperature = 11°C, and initial [As] = 5 μg L^{-1}. An example of the double concentrations peaks resulting from the push-pull effect of heat storage is indicated by the arrows in the panel for aqueous arsenic.

When the warm water is recovered, the fringe zone with elevated arsenic concentrations is pulled back towards the warm well. Consequently, two arsenic concentration peaks are observed per year in the monitoring well 5 m away from the warm ATES well (highlighted with arrows in Figure 6.5). The first arsenic peak is due to mobilization with the first warm water injection, followed by a decrease in arsenic as the water from the cold well has a lower arsenic concentration and continues passing the monitoring well. The second peak occurs when the flow reverses and arsenic that is mobilized at the fringe of the water bubble is recovered. This double peak is absent at 15 m because the maximum extension of the high arsenic concentration front approaches this monitoring well. The aqueous arsenic concentrations in the cold and warm bubbles tend to draw towards each other with progressing cycles especially near the ATES wells. At 5 m from the ATES wells, the re-distribution of sorbed arsenic, led to elevated aqueous concentrations in the cold and warm bubbles ranging between 15–20 μg L^{-1}. At 15 m from the ATES wells, the arsenic concentration in the warm bubble varies between 15–50 μg L^{-1}, but in the cold bubble arsenic concentrations are still not much beyond the initial level, showing that re-distribution of trace metals is still continuing until also here a dynamic equilibrium is achieved.

For the scenario with warm water injection at 25°C, arsenic concentrations increase near the warm well by 80%, from 5 μg L^{-1} to 12 μg L^{-1} during the first cycles (Figure S3.5). Later on, the concentrations at the centers of the cold and warm bubbles plateau at 10 μg L^{-1}, and vary between 10 and 15 μg L^{-1} at the fringe of the warm bubble. When warm water is injected at 15°C, arsenic concentrations at the warm well increase to 9 μg L^{-1} during the first cycles, after which cold and warm bubble concentrations near (5 m) the ATES wells reach a concentration of 8 μg L^{-1} (Figure S3.6). In the Netherlands, arsenic concentrations in groundwater range between 0.1–70 μg L^{-1} (Stuyfzand *et al.* 2008), and depending on the initial As concentration, and temperature rise, an increase beyond the guidance value of 10 μg L^{-1} may occur.

6.4.4 Limitations and environmental implications

Although the laboratory experiments and PHREEQC simulations extend our understanding occurring in an ATES system, various conditions/processes were not included potentially causing impacts in reality to deviate from the simulation results. First, the 1-D approach applied in the DAFT model assumes that sorbents are distributed homogeneously with depth in the aquifer. In heterogeneous aquifers, sorbents may be restricted to certain layers rich in iron oxides. Warm water injection can be efficient in leaching these layers of sorbed arsenic, but as water with elevated arsenic is recovered by the warm well and re-injected in the cold well, arsenic rich water will be distributed over the entire depth of the aquifer, also in layers which may be low in sorbents. This is expected to retard the redistribution process and may cause overall higher aqueous concentrations, especially during the first years. Second, background groundwater flow occurring at an ATES system causes the drifting away of cold and warm water bubbles, resulting in a net loss of mobilized solutes from the ATES system's zone of influence. Third, Chapter 5 showed sulfate reduction accelerates with temperatures >25°C and if sufficient iron is present iron sulfides can form which are efficient in sequestering arsenic (Beaulieu & Ramirez, 2013). In Chapter 5 it was also observed that at low temperatures (<11°C) iron-reduction becomes thermodynamically more favorable than sulfate-reduction (Bonte *et al.* 2013b, Chapter 5). This could stimulate iron-reduction in the cold bubble and sulfate-reduction in the warm bubble which in combination would result in precipitation of iron sulfides. In that case, the overall arsenic availability within the system would decrease.

Overall, the study shows that water quality field monitoring at ATES systems is required to elucidate the impact of the combinations of these processes, especially when these systems are to be realized in aquifers used for consumptive uses and already suffer from elevated concentrations of arsenic or other (oxy)anions. The development of the reactive transport model and its application to an ATES system, provides a useful tool to design monitoring programs, to minimize impacts of ATES on water quality, and to improve the conceptual understanding of thermal interactions on hydrochemistry in aquifers.

A very interesting observation made from surface complexation results and the extensive literature research is that cation sorption is generally endothermic (i.e., increasing with temperature) while (oxy)anion sorption is exothermic (i.e., decreasing with temperature). This is a crucial difference in the behavior of trace elements in aquifers subject to temperature changes, and to our knowledge not reported before in literature.

6.5 ACKNOWLEDGMENT

We acknowledge the help of Rob Stoevelaar, Rudo Verweij, Julia Claas, Mark Hanemaaijer and Valentina Chacon of VU University Amsterdam, and Sidney Meijerink and Harry van Wegen of KWR. Finally, we thank the valuable comments of three anonymous reviewers.

Chapter 7
Summary and synthesis

7.1 INTRODUCTION

In this Chapter I will summarize the outcomes and discuss what these can mean for drinking water production. Next, guidelines for water quality monitoring at shallow geothermal energy (SGE) systems will be proposed. Following this, the current policy for SGE systems in relation to groundwater protection will be reflected upon, and a suggestion is made which type of systems could be allowed in which environments without compromising the usability of groundwater as a resource for drinking water production. Last, the perspectives for future research based on the outcomes of this book will be given.

7.2 SUMMARY OF RESEARCH

7.2.1 Chapter 2: Shallow geothermal energy: A review of impacts on groundwater quality and policy in the Netherlands and European Union

In Chapter 2, a review is presented on the effects that SGE systems can impose on groundwater quality and the risks this can pose for drinking water production. Four categories of effects by SGE on groundwater quality were distinguished: (1) hydrological, (2) thermal, (3) chemical, and (4) microbial impacts. This book focuses on the latter two. A literature survey on the chemical and microbial effects of SGE showed that most published research on SGE technology focused on operational aspects such as well clogging and thermal efficiency, whereas the impacts on water quality in relation to other users, notably drinking water production, has been limited addressed in research. Particular gaps in knowledge exist on: (i) the effect of low temperature SGE systems (<25°C) based on field data, (ii) the fate of trace metals at elevated temperatures in SGE systems, and (iii) the impact of temperature changes on the microbial communities, redox processes, and hydrochemistry.

The policy survey presented in this chapter showed that regulation and research on the potential impacts of SGE on groundwater resources in many countries lags behind technological developments and the rate of deployment of this renewable energy source. This situation has improved in the Netherlands with the establishment of the Order in Council (AMvB) on SGE as of July 2013. The lack of a clear and scientifically underpinned risk management strategy implies that potentially unwanted risks might currently be taken at vulnerable locations such as near public supply well fields, whilst at other sites the application of SGE is avoided without proper reasons. Cross-sectoral subsurface planning is required to minimize negative conflicts between SGE and other subsurface interests.

7.2.2 Chapter 3: A field and modeling study of the impacts of aquifer thermal energy storage on groundwater quality

In Chapter 3 data from an aquifer thermal energy storage (ATES) system located 570 m from a public supply well field were used to investigate the impacts on groundwater quality. These data showed that groundwater circulation by the ATES system influenced groundwater quality by introducing shallow groundwater with a different chemical composition at greater depth, and vice versa. Numerical groundwater flow and solute transport modeling confirmed that this physical mixing process can explain the observed changes in groundwater quality. Because a large fraction of the water injected in the ATES wells in one season drifted away before it could be recovered in the next, it was not possible to analyze the effect of the temperature change on groundwater quality. This was the main reason to add laboratory experiments to this research.

Microbiological data showed that groundwater sampled from the ATES system contained bacteria that with standard culture tests are characterized as fecal indicator bacteria. Molecular identification however showed that most of these organisms were not related to a fecal contamination and possibly naturally present in the aquifer. Two samples contained the fecal bacteria *C. perfringens* but its presence was not linked to ATES operation and may be a relic of well construction. The presence of *C. perfringens* did not form a hygienic risk because of the sufficient travel time between the ATES wells and the public supply well field.

7.2.3 Chapter 4: Temperature-induced impacts on mobility of arsenic and other trace elements

In Chapter 4, a laboratory column testing setup was used to assess the impacts of temperature on water quality at 5°C (cold storage), 11°C (ambient temperature), 25°C (maximum allowed regular SGE), and 60°C (high temperature heat storage). Column tests were performed on anoxic, reactive, unconsolidated sandy sediments cored from a fluvial aquifer (Sterksel formation). This formation is widely used for drinking water production and SGE in the Netherlands.

The results showed that at 5°C no effects on water quality were observed compared to the reference of 11°C (in situ temperature). At 25°C, As concentrations were significantly increased and at 60°C, significant increases were observed in dissolved organic carbon (DOC), P, K, Si, As, Mo, V, B, and F concentrations. These elements should therefore be included in water quality monitoring programs of SGE projects. No consistent temperature effects were observed for Ba, Co, Cu, Ni, Pb, Zn, Eu, Ho, Sb, Sc, Yb, Ga, La, and Th, all of which were present in the sediment. The temperature-induced chemical effects were hypothesized to be caused by (incongruent) dissolution of silicate minerals (K and Si), desorption from, and potentially reductive dissolution of, iron oxides (As, B, Mo, V, and possibly P and DOC), and mineralization of sedimentary organic matter (DOC and P).

7.2.4 Chapter 5: Temperature-induced impacts on redox processes and microbial communities

Next, the column setup was used in Chapter 5 to investigate the impacts on redox processes and the associated microbial community. Both hydrochemical and molecular data showed that a temperature increase from 11°C (in-situ) to 25°C caused a shift from iron- to sulfate-reducing and methanogenic conditions. A thermodynamic analysis showed that the shift from iron- to sulfate-reducing conditions can be explained by a shift in microbial competition, with sulfate reducers becoming more efficient at utilizing substrates than iron reducers.

A further temperature increase (>45°C) resulted in the emergence of a thermophilic microbial community capable of fermentation and sulfate-reduction, but not of iron-reduction and methanogenesis. Two distinct maxima, of similar order of magnitude (5×10^{-10} M s^{-1}), were observed in sulfate-reduction rates at 40°C and 70°C. Thermophilic sulfate reduction, however, had a higher activation energy (100–160 kJ mol^{-1}) than mesophilic sulfate reduction (30–60 kJ mol^{-1}), which is probably due to a trade-off between enzyme stability and activity with thermostable enzymes being less efficient catalysts requiring higher activation energies. These results reveal that, while sulfate reducing functionality can withstand a substantial temperature rise, other key biochemical processes appear more sensitive to a temperature change.

7.2.5 Chapter 6: Reactive transport modeling of thermal column experiments to investigate the impacts of aquifer thermal energy storage on groundwater quality

In Chapter 6, the data collected with the laboratory column experiments were used to develop and calibrate a hydrogeochemical reactive transport model (PHREEQC) simulating the thermally induced water quality changes. The model included temperature-dependent surface complexation and cation-exchange, and dissolution of K-feldspar. Optimization results combined with literature data revealed that surface complexation of (oxy)anions is consistently exothermic and their sorption decreases with temperature, whereas surface complexation of cations is endothermic and cation sorption increases with temperature.

The calibrated model was applied to simulate arsenic mobility in several hypothetical aquifer thermal energy storage systems using a simple yet powerful mirrored axi-symmetrical grid. Results showed the system mobilizes arsenic towards the fringe of the warm water plume and the center of the cold water plume where these solutes become resorbed. This transient re-distribution of arsenic causes their aqueous concentrations in the cold and warm groundwater plumes to gradually approach similar elevated levels, with a final concentration depending on the average temperature over the warm and cold groundwater zones.

7.3 TRANSLATING HYDROCHEMICAL EFFECTS TO IMPACTS ON DRINKING WATER PRODUCTION

After having summarizing the principal hydrochemical effects, the next step is to discuss what these could imply for drinking water production. The first process discussed is the physical mixing of different water qualities by an ATES system changing the natural water quality zoning in the aquifer, re-distributing electron acceptors and potentially impacting on local redox conditions (Chapter 3). The effect of these processes on a drinking water pumping station depend on (the depth distribution of) the aquifer's reactivity and buffering capacity, and the well screen depth of both the pumping station and ATES system. In a worst case scenario a contaminant is brought to greater depth where sediment has a relatively high conductivity and low chemical buffering capacity. This could cause a contaminant to reach a pumping station faster and at higher concentrations. In a best case scenario, the pumping station has a well screen setting causing it already to have problems with the contaminant and the ATES system causes a dilution of the contaminant to levels below maximum allowable concentrations. Because the effects of ATES-induced mixing in the capture zone of a pumping station are likely in many cases unknown, and often highly uncertain, the establishment of ATES near a public water supply station will often interfere with the precautionary principle applied in groundwater protection zones. A complicating factor to assessing the impacts of ATES induced mixing is that the groundwater quality depth profile itself is a transient feature (Broers & van der Grift, 2004) resulting in a transient water quality at pumping stations (Mendizabal *et al.* 2012). And although the concentrations of some contaminants such as nitrate may be decreasing over recent years (Visser *et al.* 2007) the occurrence of others like pharmaceuticals may be increasing (Bound & Voulvoulis, 2005).

In Chapter 4, various species were shown to be increasingly mobile under elevated temperatures. Because most SGE systems operate at temperatures up to around 16°C, far below the maximum allowed temperature of 25°C (which was also used in our experiments), effects will be smaller for most systems. The results of surface complexation modeling combined with data from the literature (Chapter 6) suggest that sorption of (oxy)anion forming species (arsenic, boron, carbonate, phosphate, silica) is consistently exothermic and decreases with temperature, but cation sorption is endothermic (calcium, magnesium, heavy metals) and increases with temperature. This implies that for typical heavy metals, an opposite effect may be expected compared to arsenic and that their mobility decreases with temperature which was to a certain extent confirmed by Chapter 6, as sediments containing heavy metals such as lead and zinc, did not leach these elements upon a temperature increase.

A complicating factor, however, is that the combination of temperature induced desorption with mixing of groundwater by ATES, can result in mobilization of As (or other anionic trace elements) to parts of the aquifer layers with a lower sorption capacity resulting in a net shift of arsenic from the sorbed to aqueous phase. In the Netherlands, several public supply well fields abstract raw water with As concentrations above the maximum permissible concentration of drinking water (10 μg L^{-1}). These levels (up to 70 μg L^{-1}) are in general easily removed through rapid sand filtration thanks to accompanying high Fe^{2+} concentrations (Stuyfzand *et al.* 2006). When however arsenic desorbs due to a temperature increase, the Fe^{2+} concentration will not increase along with it, unless reductive dissolution of iron oxides occurs as well. This combined behavior was shown in Chapter 5, but occurred only at the lower temperatures for one of the sediments tested. If iron-reduction occurs concomitantly with sulfate-reduction, precipitation of iron-sulfides may also sequester arsenic that is thermally desorbed. The combination of stable iron oxides and thermally desorbing arsenic is thus possibly more problematic, and could necessitate an additional treatment step, for example, coagulation with $FeCl_3$ (Hering *et al.* 1997). For heavy metals, the effect could again be the opposite, leading to an increase in net sorption in the aquifer and an improvement in water quality.

A second water quality parameter observed to increase with temperature, was the DOC concentration. This can cause color problems in drinking water which may necessitate a discoloration treatment step (such as oxidation) of the water abstracted (Wallage *et al.* 2006). An advanced oxidation step can however cause problems with toxic disinfection by-products (Von Gunten, 2003). An important notion that was inferred in Chapter 5 and also observed elsewhere (Evans *et al.* 2005; Xu & Saiers, 2010) is, that DOC mobilized at higher temperatures is relatively recalcitrant and thus less easily removed in water treatment. On the up-side, it is also less likely to cause well clogging in SGE systems. A final relevant aspect to the increasing DOC concentrations is that humic substances constituting the bulk of DOC can form complexes with trace elements, causing them to remain mobile and not to become resorbed with decreasing temperature outside the direct zone of influence of the ATES system (Cao *et al.* 1995).

The increasing mineralization rate of sedimentary organic matter (SOM) at temperatures >25°C is likely to lead to a reduction of the buffering capacity of the aquifer. SOM is particularly important for the sorption and retardation of hydrophobic organic micro-pollutants (Appelo & Postma, 2005). The associated promotion of sulfate-reduction following the fermentation of SOM described in Chapter 5 however, can lead to the formation of iron sulfides, which sequester thermally mobilized anionic trace elements. This has a positive impact on groundwater quality (and is in fact also used for remediation of groundwater with high arsenic concentrations (Beaulieu & Ramirez, 2013).

7.4 GROUNDWATER QUALITY MONITORING NEAR SGE SYSTEMS

A next step is to derive meaningful monitoring requirements for those locations where SGE systems are planned at some distance from sites where groundwater is extracted for other uses. Table 7.1 provides a comprehensive proposal for monitoring at these sites, and provides a key to the relevant processes and sections in this book which address them. Also shown is a number of monitoring parameters, required for operational aspects to complete this overview. A site dependent selection can be made from this table, based on vulnerability of other groundwater uses and anticipated relevant processes.

Table 7.1 Groundwater quality monitoring guidelines for SGE systems.

Group	Parameter	Reason and further reference
Monitoring groundwater quality in relation to other uses		
Field parameters	pH, Electrical conductivity, temperature (down hole), dissolved oxygen (DO)	Monitor mixing of different water qualities and thermal impacts (Chapter 3).
Macro parameters	Cl, SO_4, alkalinity, NO_3, NH_4, Ca, Mg, Na, K	Quality assurance of analyses (electrical balance, also used for many other aspects discussed below).
(Toxic) trace elements	As, Mo, V, B, Cr, Be and F or a complete ICP-MS scan	Mobilization due to desorption or reductive dissolution (Chapter 4).
Redox parameters	Fe, SO_4, NO_3, NH_4, CH_4, DOC, PO_4	Monitor changing redox processes (Chapter 5).
Contaminants present in ATES influence area	Depends on type of contaminants present. Broad standard packages can be used such PAH, BTEX, VOCl[a]	Monitor spreading of contaminants by increased flow velocities (briefly discussed in Chapter 2, and Zuurbier *et al.* (2013b)
Pathogens	*E-coli* + molecular confirmation	Assure system is not open to microbial contamination
Operational aspects		
Carbonate scaling	Ca, Mg, alkalinity, pH, temp, Fe, DOC	Prevent precipitation of carbonates ($T > 30°C$) (Griffioen & Appelo, 1993)
Iron precipitation	Fe, Mn, pH, temp, DO and NO_3	Prevent iron precipitation (Houben & Trestakis, 2007)
Iron sulphides scaling	Fe, SO_4, pH, HS^-	Prevent iron sulfide scaling due to SO_4 reducing conditions (briefly discussed in Chapter 5, see also (Brons, 1992))
Gas pressure	Partial pressures of CH_4, N_2, CO_2	Prevent degassing and gas bubble clogging or intrusion of oxygen
Corrosion	SO_4, Cl	Prevent anaerobic corrosion (briefly discussed in Chapter 5, see (Brons, 1992; Lerm *et al.* 2013)

[a]PAH, polycyclic aromatic hydrocarbons; BTEX, benzene toluene ethylbenzene and xylene; VOCl, Volatile Chlorinated Organic Compounds.

The monitoring locations can be a filter set in the annulus of the ATES well, a tap point of the ATES wells, or a dedicated monitoring well located at some distance from the ATES well or BTES borehole. Note that for ATES systems, it is important to have a good understanding of the dynamics of the systems. Common monitoring pitfalls are:

- When monitoring water quality at the 'warm' well at the end of summer, one may expect to sample the maximum effect of heat storage. The warm well at the end of summer however contains recently injected water from the cold well, and if there is some bell drift in the cold bubble, it will contain groundwater hardly influenced by ATES. Timing of sampling events of groundwater in the ATES well, should thus be based on the warm or cold water bubble size, the velocity of bell drift, and the flow through the system.
- The redistribution of trace elements in an ATES system and resorption should be considered when monitoring for these solutes. Modeling in Chapter 6 showed that with progressing cycles, concentrations in cold and warm wells will become similar and a reference well outside the zone of influence should be used to determine the net effect of the system.
- Locations of monitoring wells are often based on the anticipated design circulation volume of the ATES system which are often much smaller than actual volumes. This causes monitoring wells to be situated outside the zone of influence.
- ATES system temperature readings should be used with caution as sensors may be located above ground. The best method is to take a submersible temperature sensor and take readings at the depth of the filter from where the groundwater sample is taken.

7.5 POLICY PERSPECTIVES

With the recently introduced legislation on shallow geothermal energy (the Order in Council on shallow geothermal energy, here abbreviated as the AMvB, corresponding to the generally used Dutch abbreviation), a new legal framework is introduced in the Netherlands as of July 2013. The rules for groundwater protection zones are not addressed in this AMvB and remain secured under the national Environment protection act (abbreviated under the generally used term Wm). The Wm directs provincial authorities to define groundwater protection areas in which rules apply to protect groundwater quality for drinking water production.

An improvement for groundwater protection in a broader sense (so outside groundwater protection zones) with the introduction of the AMvB SGE, is that both closed and open loop systems are now required to be registered with the authorities (respectively municipalities and provinces). Furthermore, a training and certification system is established to educate drilling companies aiming to assure high quality work be delivered which also strengthens groundwater protection, for example by strictly following guidelines on resealing of perforated aquitards. Regulatory enforcement by the responsible authorities will be required to secure that these protocols are followed. The effort required to get the permit for an ATES system has decreased as the previously required, extensive impact assessment report is no longer required and the processing time has decreased from >6 months to 8 weeks. In many areas in the Netherlands (especially the west), ATES systems are realized in deep saline aquifers where historically no other users are present, justifying this less restrictive legislation. A complicating factor here is however that the horticultural sector in this area is increasingly using the subsurface for storage of fresh water (Zuurbier *et al.* 2013a). This highlights the need to broaden policies dealing with the subsurface balancing interests of all subsurface users as discussed in Chapter 2.

An important aspect of the new SGE legislative framework is that it explicitly gives local authorities the possibility for allowing higher maximum re-injection temperatures than the default threshold levels of 25°C and 30°C for open and closed loop systems, respectively. Also, the provincial authorities can decide on whether SGE could be allowed in groundwater protection zones. The possibility of higher temperatures was already present in previous legislation (the Water Act), but the AMvB specifically states that a temperature above the 'default' maximum injection temperature is allowed and can result in higher efficiencies of systems and many provinces are now investigating how and where to allow high temperature heat storage (e.g., the province of Gelderland as discussed in Zaadnoordijk *et al.* (2013)). The decision on whether SGE could be allowed within groundwater protection zones also basically remains with the provincial authorities who regulate groundwater protection zones. It is noted that there is not a strict prohibition of SGE within protection zones (which already led to a number of ATES systems in groundwater protection zones shown in Figure 1.3). On the basis of the results of this book, a suggestion is made how these rules could be interpreted for different groundwater regions in the Netherlands (Table 7.2).

Table 7.2 Policy on SGE in the Netherlands: current situation and perspectives for refinements.

Region	Current relevant legislation following from the AMvB and Wm	Suggested policy interpretation
I. Groundwater abstraction region (generally 60 day zone)	Generally not allowed (often in ownership of water utility) (enforced under the Environmental protection act)	Precautionary principle prevails.
II. Groundwater protection zone (generally 25 year travel zone)	Generally not allowed (enforced under the Environmental protection act)	Precautionary principle prevails. Possible exceptions are well fields that are to be closed.
III. Capture zone, but outside protection zone	<25°C re-injection temperature (open systems) and <30°C for the circulation fluids (closed systems)	Authority may consult water utility or other consumptive water users. In case of vulnerable well fields, the water utility could protest against approval of the SGE system or could suggest it to operate at a reduced temperature regime.
IV. Other fresh groundwater bodies	Higher temperatures can be possible when this does not interfere with other uses.	The requirements in the AMvB SGE are sufficient to address possible interferences with other groundwater users.
V. Shallow saline groundwater bodies	A heat balance is required (net cooling is sometimes allowed)	High temperature heat storage is not advisable as temperature effects can propagate into surface water bodies.
VI. Deep saline groundwater bodies		In these groundwater bodies high temperature heat storage could be allowed provided it does not increase the temperature in adjacent fresh aquifers.

In regions where restrictions apply for drinking water protection (Region I and II, Table 7.2), the precautionary principle should generally prevail. Only in extraordinary cases an SGE system should be feasible. An example could be a planned closure of a pumping station, or if the benefits of allowing SGE are such that it would compensate for potential adverse effects for the water supply. It is unlikely that the benefits of an individual SGE system (for a building or group of buildings) could outweigh the interest of water supply of a city or region. So this option could only be feasible in a regional context where heat and water supply are considered together.

Under the current rules, SGE systems can be established outside the protection zone, which can still be part of the capture zone of a PSWF (Region III). In these cases, it would be desirable to have a proper impact assessment. Currently, applications for SGE systems are handled whenever a license application arrives at the regulating authority. And given the shortened handling times for license approvals, it may be become hard to carry out a proper impact assessment. A more thorough risk assessment in the capture zones may be possible if an ex ante evaluation of the impacts of SGE in the capture zone are combined with the broader water quality risk assessment carried out for the so-called protected area dossier (Gebiedsdossiers), a tool resulting from the EU Water Framework directive for protection of drinking water resources. Currently, this dossier mainly focuses on water quality threats from land surface (fertilizers, pesticides, sewers and point or diffuse pollution plumes).

The application of higher temperature SGE seems only warranted in deep saline aquifers (Region VI). Saline aquifers do not provide groundwater that can easily be used for drinking water production, at least not using a 'basic water treatment', as described in the EU Water Framework directive. It is noted that the definition of a 'basic' water treatment differs regionally and by source water, but for groundwater aeration, rapid sand filtration and, if required, a disinfection step can be considered as basic steps (Wuijts & Rijswick, 2007). Shallow saline groundwater bodies (zone V) are less suitable as the impacts of higher temperature SGE can more easily dissipate into surface waters, and thus form a risk for aquatic ecology and recreational use of these waters.

7.6 RESEARCH PERSPECTIVES

The last part of the book provides a short overview of relevant new research perspectives that follow from the results:

- The strong temperature dependence of arsenic (and other oxyanions) mobility, also in the range up to 25°C, is important to consider especially when SGE is realized in aquifers used for drinking water production. A crucial next step is to validate the results of this study in field SGE systems (operating at 25°C or higher) in the Sterksel formation and to investigate the potential arsenic mobilization in other geological formations. It would be good to derive a more generic relationship between the geochemical reactivity of the Dutch geological formations (Gaans *et al.* 2011; Griffioen *et al.* 2012) and the response of these to temperature increases. This would allow an efficient scaling up of the results found here, and facilitate an effective monitoring in sensitive areas.
- In this book, I have described the mixing effect that an ATES may have on the distribution of electron acceptors and redox processes in an aquifer on the one hand, and the thermal effects on trace element mobility and redox processes the other. I have described the combination of these two processes qualitatively in several chapters, but investigating this interaction in a more quantitative way is required to get a more thorough understanding of the impacts of ATES on groundwater quality. This could be achieved by detailed field monitoring, especially at some of the high temperature heat storage systems that are to be realized in the coming years, combined with reactive transport modeling. The reactive transport modeling framework presented in Chapter 6 can provide a good starting point for this, but up-scaling to a full 3D model is required to assess the cumulative effects on downstream groundwater quality and impacts to other groundwater uses.

- The temperature dependence of sorption might also provide opportunities to optimize in- or ex-situ arsenic remediation systems. Arsenic is a common issue in aquifer storage and recovery (ASR) systems, where oxidation of pyrite causes the mobilization of arsenic (Wallis *et al.* 2011; Mirecki *et al.* 2012). During the storage phase of ASR, the arsenic concentration in the ASR injected water bubble is controlled by the equilibrium with arsenic sorbed to neo-formed iron hydroxides (Wallis *et al.* 2011). Combining heat storage with ASR could be a method of accelerating the removal of sorbed arsenic following pyrite oxidation and pre-conditioning the aquifer for storage of raw or drinking water. Cold water (having a relatively high dissolved oxygen concentration) effectively oxidizes pyrite releasing arsenic which is sorbed to neo-formed iron hydroxides. In the next season relatively warm water is injected which flushes arsenic from the sorbent. This causes an initial rapid flush with arsenic (which may be treated or disposed of), and not a long tail of elevated arsenic concentrations.
- Given the explicit policy freedom for high temperature heat storage, aiming to boost application of this technology, a wider understanding of the dynamics of thermophilic sulfate reducing bacterial communities should be developed. It was shown in this book that these communities can develop in Dutch sediments, and in Germany Lerm *et al.* (2013) demonstrated that these can cause damage to these systems by anaerobic corrosion.

References

Altschul S. F., Gish W., Miller W., Myers E. W. and Lipman D. J. (1990). Basic local alignment search tool. *Journal of Molecular Biology*, **215**(3), 403–410.

Amy G. L., Sierka R. A., Bedessem J., Price D. and Tan L. (1992). Molecular size distributions of dissolved organic matter. *Journal/American Water Works Association*, **84**(6), 67–75.

Angove M. J., Wells J. D. and Johnson B. B. (1999). The influence of temperature on the adsorption of Cadmium(II) and Cobalt(II) on goethite. *Journal of Colloid and Interface Science*, **211**(2), 281–290.

Anonymous (1998). On the Quality of Water Intended for Human Consumption. Council directive 98/83/EC, European Union, Brussels.

Anonymous (2012). SciPy Reference Guide Release 0.11.0.dev-659017f. p. 969. http://docs.scipy.org/doc/scipy/scipy-ref.pdf (accessed 1 July 2013)

Appelo C. A. J. (1994). Cation and proton exchange, pH variations, and carbonate reactions in a freshening aquifer. *Water Resource Research* **30**(10), 2793–2805.

Appelo C. A. J. and Postma D. (2005). Geochemistry, Groundwater and Pollution, 2nd edn. A.A. Balkema, Leiden, p. 649.

Appelo C. A. J., Griffioen J. and van der Weiden M. J. J. (1990). Environmental and Chemical Aspects of Thermal Energy Storage in Aquifers and Research and Development of Water Treatment Methods: Subtask B: (Bio) geochemical reactions. Vrije Universiteit, Amsterdam, p. 115.

Appelo C. A. J., Van Der Weiden M. J. J., Tournassat C. and Charlet L. (2002). Surface complexation of ferrous iron and carbonate on ferrihydrite and the mobilization of arsenic. *Environmental Science & Technology*, **36**(14), 3096–3103.

Arning E., Kölling M., Schulz H. D., Panteleit B. and Reichling J. (2006a). Einfluss oberflächennaher Wärmegewinnung auf geochemische Prozesse im Grundwasserleiter. *Grundwasser*, **11**(1), 27–39.

Arning E., Kölling M., Schulz H. D., Panteleit B. and Reichling J. (2006b). The influence of shallow geothermal energy on geochemical processes in aquifers (in German). *Grundwasser*, **11**(1), 27–39.

Arnosti C., Jørgensen B. B., Sagemann J. and Thamdrup B. (1998). Temperature dependence of microbial degradation of organic matter in marine sediments: polysaccharide hydrolysis, oxygen consumption, and sulfate reduction. *Marine Ecology Progress Series*, **165**, 59–70.

Avci C. B. (1992). Flow occurrence between confined aquifers through improperly plugged boreholes. *Journal of Hydrology*, **139**(1–4), 97–114.

Bae S. S., Lee J.-H. and Kim S.-J. (2005). Bacillus alveayuensis sp. nov., a thermophilic bacterium isolated from deep-sea sediments of the Ayu Trough. *International Journal of Systematic and Evolutionary Microbiology*, **55**(3), 1211–1215.

Baes C. F. Jr and Mesmer R. F. (1981). The thermodynamics of cation hydrolysis. *American Journal of Science*, **281**(7), 935–962.

Bakr M., van Oostrom N. and Sommer W. (2013). Efficiency of and interference among multiple aquifer thermal energy storage systems; A Dutch case study. *Renewable Energy*, **60**(0), 53–62.

Banerjee K., Amy G. L., Prevost M., Nour S., Jekel M., Gallagher P. M. and Blumenschein C. D. (2008). Kinetic and thermodynamic aspects of adsorption of arsenic onto granular ferric hydroxide (GFH). *Water Research*, **42**(13), 3371–3378.

Bauer M. and Blodau C. (2006). Mobilization of arsenic by dissolved organic matter from iron oxides, soils and sediments. *Science of The Total Environment*, **354**(2–3), 179–190.

Bayer P., Saner D., Bolay S., Rybach L. and Blum P. (2012). Greenhouse gas emission savings of ground source heat pump systems in Europe: a review. *Renewable and Sustainable Energy Reviews*, **16**(2), 1256–1267.

Beaulieu B. and Ramirez R. E. (2013). Arsenic remediation field study using a sulfate reduction and zero-valent iron PRB. *Groundwater Monitoring & Remediation*, **33**(2), 85–94.

Benner S. G., Blowes D. W., Ptacek C. J. and Mayer K. U. (2002). Rates of sulfate reduction and metal sulfide precipitation in a permeable reactive barrier. *Applied Geochemistry*, **17**(3), 301–320.

Bense V. F. and Van Balen R. (2004). The effect of fault relay and clay smearing on groundwater flow patterns in the lower rhine embayment. *Basin Research*, **16**(3), 397–411.

Bense V. F., Van den Berg E. H. and Van Balen R. T. (2003). Deformation mechanisms and hydraulic properties of fault zones in unconsolidated sediments; the Roer Valley Rift System, The Netherlands. *Hydrogeology Journal*, **11**(3), 319–332.

Bisson J. W. and Cabelli V. J. (1980). Clostridium perfringens as a water pollution indicator. *Journal of the Water Pollution Control Federation*, **52**(2), 241–248.

Blum P., Campillo G., Münch W. and Kölbel T. (2010). CO_2 savings of ground source heat pump systems – A regional analysis. *Renewable Energy*, **35**(1), 122–127.

Bobylev N. (2009). Mainstreaming sustainable development into a city's Master plan: a case of Urban Underground Space use. *Land Use Policy*, **26**(4), 1128–1137.

Bond W. J. and Phillips I. R. (1990). Cation exchange isotherms obtained with batch and miscible-displacement techniques. *Soil Science Soceity American Journal*, **54**(3), 722–728.

Bonte M. (2013). Impacts of Shallow Geothermal Energy On Groundwater Quality, Ph.D. Dissertation VU Amsterdam, Amsterdam.

Bonte M., Van Den Berg G. and Wezel A. M. (2008). Underground thermal energy storage in relation to groundwater protection (in Dutch). *Bodem*, 5/2008, 22–26.

Bonte M., Stuyfzand P. J., Hulsmann A. and Van Beelen P. (2011a). Underground thermal energy storage: Environmental risks and policy developments in the Netherlands and the EU. *Ecology and Society*, **16**(1), 22.

Bonte M., Stuyfzand P. J. and Van Den Berg G. (2011b). The effects of aquifer thermal energy storage on groundwater quality and the consequences for drinking water production: a case study from the Netherlands. *Water Science and Technology*, **63**(9), 1922–1931.

Bonte M., van Breukelen B. M. and Stuyfzand P. J. (2013a). Temperature-induced impacts on groundwater quality and arsenic mobility in anoxic aquifer sediments used for both drinking water and shallow geothermal energy production. *Water Research*, **47**(14), 5088–5100.

Bonte M., Röling W. F. M., Zaura E., van der Wielen P. W. J. J., Stuyfzand P. J. and van Breukelen B. M. (2013b). Impacts of shallow geothermal energy production on redox processes and the microbial community. *Environmental Science & Technology*, 47(24), 14476–14484.

Boujelben N., Bouzid J. and Elouear Z. (2009). Adsorption of nickel and copper onto natural iron oxide-coated sand from aqueous solutions: study in single and binary systems. *Journal of Hazardous Materials*, **163**(1), 376–382.

Bound J. P. and Voulvoulis N. (2005). Household disposal of pharmaceuticals as a pathway for aquatic contamination in the United Kingdom. *Environmental Health Perspectives*, **113**(12), 1705–1711.

Brandt B. W., Bonder M. J., Huse S. M. and Zaura E. (2012). TaxMan: a server to trim rRNA reference databases and inspect taxonomic coverage. *Nucleic Acids Research*, **40**(W1), W82–W87.

Breeuwsma A. (1987). The mineralogical composition of Dutch sand and clay soil. In Soil Science of the Netherlands, W. P. Locher and H. d. Bakker (eds), Ministry of Agriculture and Fishery, Den Bosch, pp. 95–99, Chapter 7 (in Dutch).

Brielmann H., Griebler C., Schmidt S. I., Michel R. and Lueders T. (2009). Effects of thermal energy discharge on shallow groundwater ecosystems. *Fems Microbiology Ecology*, **68**(3), 273–286.

Brielmann H., Lueders T., Schreglmann K., Ferraro F., Avramov M., Hammerl V., Blum P., Bayer P. and Griebler, C. (2011). Shallow geothermal energy usage and its potential impacts on groundwater ecosystems (Oberflächennahe Geothermie und ihre potenziellen Auswirkungen auf Grundwasserökosysteme). *Grundwasser*, **16**(2), 77–91.

Brigante M., Zanini G. and Avena M. (2007). On the dissolution kinetics of humic acid particles: effects of pH, temperature and Ca2+ concentration. *Colloids and Surfaces A: Physicochemical and Engineering Aspects*, **294**(1–3), 64–70.

Broers H. P. and van der Grift B. (2004). Regional monitoring of temporal changes in groundwater quality. *Journal of Hydrology*, **296**(1–4), 192–220.

Brons H. J. (1992). Biogeochemical Aspects of Aquifer Thermal Energy Storage, Brons. PhD thesis Wageningen University, Wageningen. http://edepot.wur.nl/201141 (accessed 1 July 2013)

Brons H. J. and Zehnder A. J. B. (eds) (1990). Biochemical Aspects of Aquifer Thermal Energy Storage. TNO, The Hague.

Brons H. J., Griffioen J., Appelo C. A. J. and Zehnder A. J. B. (1991). (Bio)geochemical reactions in aquifer material from a thermal energy storage site. *Water Research*, **25**(6), 729–736.

Brun A. and Engesgaard P. (2002). Modelling of transport and biogeochemical processes in pollution plumes: literature review and model development. *Journal of Hydrology*, **256**(3–4), 211–227.

Brunauer S., Emmett P. H. and Teller E. (1938). Adsorption of gases in multimolecular layers. *Journal of the American Chemical Society*, **60**(2), 309–319.

Caljé R. J. (2010). Future use of Aquifer Thermal Energy Storage below the historic centre of Amsterdam, TU Delft, Masters, Delft. http://www.citg.tudelft.nl/fileadmin/Faculteit/CiTG/Over_de_faculteit/Afdelingen/Afdeling_watermanagement/Secties/waterhuishouding/Leerstoelen/Hydrologie/Education/MSc/Past/doc/Calje_2010.pdf (accessed 1 July 2013)

Cao Y., Conklin M. and Betterton E. (1995). Competitive complexation of trace metals with dissolved humic acid. *Environmental Health Perspectives*, **103**(Suppl. 1), 29–32.

Caporaso J. G., Bittinger K., Bushman F. D., Desantis T. Z., Andersen, G. L. and Knight, R. (2009). PyNAST: a flexible tool for aligning sequences to a template alignment. *Bioinformatics*, **26**(2), 266–267.

Caporaso J. G., Kuczynski J., Stombaugh J., Bittinger K., Bushman F. D., Costello E. K., Fierer N., Pẽa A. G., Goodrich J. K., Gordon J. I., Huttley G. A., Kelley S. T., Knights D., Koenig J. E., Ley R. E., Lozupone C. A., McDonald D., Muegge B. D., Pirrung M., Reeder J., Sevinsky J. R., Turnbaugh P. J., Walters W. A., Widmann J., Yatsunenko T., Zaneveld J. and Knight R. (2010). QIIME allows analysis of high-throughput community sequencing data. *Nature Methods*, **7**(5), 335–336.

CBS (2008). Sustainable Energy in the Netherlands 2007. Centraal Bureau voor de Statistiek, Rijswijk (in Dutch).

Chambon J. C., Bjerg P. L., Scheutz C., Baelum J., Jakobsen R. and Binning P. J. (2013). Review of reactive kinetic models describing reductive dechlorination of chlorinated ethenes in soil and groundwater. *Biotechnology and Bioengineering*, **110**(1), 1–23.

Chesnaux R., Chapuis R. P. and Molson J. W. (2006). A new method to characterize hydraulic short-circuits in defective borehole seals. *Ground Water*, **44**(5), 676–681.

Cole J. R., Wang Q., Cardenas E., Fish J., Chai B., Farris R. J., Kulam-Syed-Mohideen A. S., McGarrell D. M., Marsh T., Garrity G. M. and Tiedje J. M. (2009). The Ribosomal Database Project: improved alignments and new tools for rRNA analysis. *Nucleic Acids Research*, **37**(Suppl. 1), D141–D145.

Compendium voor de leefomgeving Compendium voor de leefomgeving: Veranderingen bodemgebruik, 1979–2008. http://www.compendiumvoordeleefomgeving.nl/indicatoren/nl0060-Bodemgebruik-in-Nederland.html?i=15-18 (accessed 30 June 2013).

Cozzarelli I. M., Bekins B. A., Baedecker M. J., Aiken G. R., Eganhouse R. P. and Tuccillo M. E. (2001). Progression of natural attenuation processes at a crude-oil spill site: 1. Geochemical evolution of the plume. *Journal of Contaminant Hydrology*, **53**(3–4), 369–385.

D'Amico S., Marx J.-C., Gerday C. and Feller G. (2003). Activity-stability relationships in extremophilic enzymes. *Journal of Biological Chemistry*, **278**(10), 7891–7896.

Dai Z. and Samper J. (2004). Inverse problem of multicomponent reactive chemical transport in porous media: formulation and applications. *Water Resource Research*, **40**(7), W07407.

Danson M. J., Hough D. W., Russell R. J. M., Taylor G. L. and Pearl L. (1996). Enzyme thermostability and thermoactivity. *Protein Engineering*, **9**(8), 629–630.

Davidson E. A. and Janssens I. A. (2006). Temperature sensitivity of soil carbon decomposition and feedbacks to climate change. *Nature*, **440**(7081), 165–173.

De Bussetti S. G., Ferreiro E. A. and Helmy A. K. (1995). Sorption of boron by hydrous Al-oxide. *Clays and Clay Minerals*, **43**(1), 58–62.

De Vries J. J. (2007). Groundwater. In: Geology of the Netherlands, T. E. Wong, D. A. J. Batjes and J. d. Jager (eds), Royal Netherlands Academy of Arts and Sciences, Amsterdam, pp. 295–315.

DeSutter T. M., Pierzynski G. M. and Baker L. R. (2006). Flow-through and batch methods for determining calcium-magnesium and magnesium-calcium selectivity contribution no. 04-251-J from the kansas agric. exp. stn., Manhattan, KS. *Soil Science Soceity American Journal*, **70**(2), 550–554.

Dinkla I., Lieten S., Drijver B. and Hartog N. (2012). Effects of aquifer thermal energy storage on geochemistry and biology, Final Report Monitoring and Laboratory Testing (in Dutch), IF technology, p. 85.

Dixit S. and Hering J. G. (2003). Comparison of Arsenic(V) and Arsenic(III) sorption onto iron oxide minerals: implications for arsenic mobility. *Environmental Science & Technology*, **37**(18), 4182–4189.

Doherty J. (2010). PEST, Model-Independent Parameter Estimation, 5th edn. User Manual, p. 336.

Drijver B. J. and Wennekes R. G. A. (2013). Onderzoek Naar Interferentie Tussen Open en Gesloten Bodemenergiesystemen. IF-technology, Arnhem, p. 46.

Driscoll F. G. (1986). Groundwater and Wells Johnson Filtration Systems. St. Paul, Minnesota, p. 1021.

Dufour F. C. (2000). Groundwater in the Netherlands – Facts and Figures. TNO Geosciences, Utrecht, p. 96.

Dzombak D. A. and Morel F. M. M. (1990). Surface Complexation Modeling: Hydrous Ferric Oxide. John Wiley & Sons, New York, p. 393.

Edgar R. C. (2010). Search and clustering orders of magnitude faster than BLAST. *Bioinformatics,* **26**(19), 2460–2461.

Edgar R. C., Haas B. J., Clemente J. C., Quince C. and Knight R. (2011). UCHIME improves sensitivity and speed of chimera detection. *Bioinformatics*, **27**(16), 2194–2200.

Elsgaard L., Prieur D., Mukwaya G. M. and Jørgensen B. B. (1994). Thermophilic sulfate reduction in hydrothermal sediment of lake tanganyika, East Africa. *Applied and Environmental Microbiology*, **60**(5), 1473–1480.

Engineering toolbox http://www.engineeringtoolbox.com/ (accessed 19 March 2013).

Eugster W. J. and Sanner B. (2007). Technological status of shallow geothermal energy in Europe, European Geothermal Congress 2007, Unterhaching Germany, 30 May – 1 June 2007, Unterhaching Germany.

Evanko C. R. and Dzombak D. A. (1998). Influence of structural features on sorption of NOM-analogue organic acids to goethite. *Environmental Science & Technology*, **32**(19), 2846–2855.

Evans C. D., Monteith D. T. and Cooper D. M. (2005). Long-term increases in surface water dissolved organic carbon: observations, possible causes and environmental impacts. *Environmental Pollution*, **137**(1), 55–71.

Evans D., Stephenson M. and Shaw R. (2009). The present and future use of 'land' below ground. *Land Use Policy*, **26**(Suppl. 1), S302–S316.

Feller G. (2003). Molecular adaptations to cold in psychrophilic enzymes. *Cellular and Molecular Life Sciences*, **60**(4), 648–662.

Ferguson G. (2006). Potential use of particle tracking in the analysis of low-temperature geothermal developments. *Geothermics*, **35**(1), 44–58.

Ferguson G. (2009). Unfinished business in geothermal energy. *Ground Water*, **47**(2), 167–167.

Ferguson G. and Woodbury A. D. (2006). Observed thermal pollution and post-development simulations of low-temperature geothermal systems in Winnipeg, Canada. *Hydrogeology Journal*, **14**(7), 1206–1215.

Ferguson G. and Woodbury A. D. (2007). Urban heat island in the subsurface. *Geophysical Research Letters*, **34**(23), 4.

Filius J. D., Lumsdon D. G., Meeussen J. C. L., Hiemstra T. and Van Riemsdijk W. H. (2000). Adsorption of fulvic acid on goethite. *Geochimica et Cosmochimica Acta*, **64**(1), 51–60.

Fletcher K. E., Costanza J., Pennell K. D. and Löffler F. E. (2011). Electron donor availability for microbial reductive processes following thermal treatment. *Water Research*, **45**(20), 6625–6636.

Fokkink L. G. J., De Keizer A. and Lyklema J. (1990). Temperature dependence of cadmium adsorption on oxides: I. Experimental observations and model analysis. *Journal of Colloid and Interface Science*, **135**(1), 118–131.

Fokkink L. G. J., Rhebergen A. G., de Keizer A. and Lyklema J. (1992). Temperature dependence of cadmium adsorption on oxides: Part 2. Thermodynamics and microcalorimetry of the Cd(NO3)2 rutile system. *Journal of Electroanalytical Chemistry*, **329**(1–2), 187–199.

Freedman V. L., Waichler S. R., Mackley R. D. and Horner J. A. (2012). Assessing the thermal environmental impacts of an groundwater heat pump in southeastern Washington State. *Geothermics*, **42**(0), 65–77.

Friis A. K., Albrechtsen H. J., Cox E. and Bjerg P. L. (2006a) The need for bioaugmentation after thermal treatment of a TCE-contaminated aquifer: laboratory experiments. *Journal of Contaminant Hydrology*, **88**(3–4), 235–248.

Friis A. K., Heron G., Albrechtsen H. J., Udell K. S. and Bjerg P. L. (2006b) Anaerobic dechlorination and redox activities after full-scale Electrical Resistance Heating (ERH) of a TCE-contaminated aquifer. *Journal of Contaminant Hydrology*, **88**(3–4), 219–234.

Gaans P. M., Griffioen J., Mol G. and Klaver G. (2011). Geochemical reactivity of subsurface sediments as potential buffer to anthropogenic inputs: a strategy for regional characterization in the Netherlands. *Journal of Soils and Sediments*, **11**(2), 336–351.

Gao Q., Li M., Yu M., Spitler J. D. and Yan Y. Y. (2009). Review of development from GSHP to UTES in China and other countries. *Renewable and Sustainable Energy Reviews*, **13**(6–7), 1383–1394.

Goldberg S. and Johnston C. T. (2001). Mechanisms of arsenic adsorption on amorphous oxides evaluated using macroscopic measurements, vibrational spectroscopy, and surface complexation modeling. *Journal of Colloid and Interface Science*, **234**(1), 204–216.

Goldberg S., Forster H. S. and Heick E. L. (1993). Temperature effects on boron adsorption by reference minerals and soils. *Soil Science*, **156**(5), 316–321.

Goldscheider N., Hunkeler D. and Rossi P. (2006). Review: microbial biocenoses in pristine aquifers and an assessment of investigative methods. *Hydrogeology Journal*, **14**(6), 926–941.

Gräfe M., Mustafa G., Singh B. and Kookana R. S. (2007). Chapter 7 Temperature and Aging Effects on the Surface Speciation of Cd(II) at the Goethite–Water Interface. In: Developments in Earth and Environmental Sciences, O. B. Mark and B. K. Douglas (eds), Elsevier, St Louis, Vol. 7, pp. 187–204.

Greskowiak J., Prommer H., Massmann G. and Nützmann G. (2006). Modeling seasonal redox dynamics and the corresponding fate of the pharmaceutical residue phenazone during artificial recharge of groundwater. *Environmental Science & Technology*, **40**(21), 6615–6621.

Griebler C. and Lueders T. (2009). Microbial biodiversity in groundwater ecosystems. *Freshwater Biology*, **54**(4), 649–677.

Griffioen J. and Appelo C. A. J. (1993). Nature and extent of carbonate precipitation during aquifer thermal energy storage. *Applied Geochemistry*, **8**(2), 161–176.

Griffioen J., Appelo C. A. J. and Van Veldhuizen M. (1992). Practice of chromatography: deriving isotherms from elution curves. *Soil Science Society of America Journal*, **56**(5), 1429–1437.

Griffioen J., Klein J. and van Gaans P. F. M. (2012). Reaction capacity characterization of shallow sedimentary deposits in geologically different regions of the Netherlands. *Journal of Contaminant Hydrology*, **127**(1–4), 30–46.

Hähnlein S., Bayer P. and Blum P. (2010a). International legal status of the use of shallow geothermal energy. *Renewable & Sustainable Energy Reviews*, **14**(9), 2611–2625.

Hähnlein S., Molina-Giraldo N., Blum P., Bayer P. and Grathwohl P. (2010b). Ausbreitung von Kältefahnen im Grundwasser bei Erdwärmesonden. *Grundwasser*, **15**(2), 123–133.

Hähnlein S., Bayer P., Ferguson G. and Blum P. (2013). Sustainability and policy for the thermal use of shallow geothermal energy. *Energy Policy*, **59**, 914–925.

Hall E. K., Neuhauser C. and Cotner J. B. (2008). Toward a mechanistic understanding of how natural bacterial communities respond to changes in temperature in aquatic ecosystems. *ISME Journal*, **2**(5), 471–481.

Hammer Ø., Harper D. A. T. and Ryan P. D. (2001). Past: paleontological statistics software package for education and data analysis. *Palaeontologia Electronica*, **4**(1), http://palaeo-electronica.org/2001_1/past/issue1_01.htm

Hansen H. C. B., Wetche T. P., Raulund-Rasmussen K. and Borggaard O. K. (1994). Stability constants for silicate adsorbed to ferrihydrite. *Clay Minerals*, **29**(3), 341–350.

Hansen J. W., Thamdrup B. and Jørgensen B. B. (2000). Anoxic incubation of sediment in gas-tight plastic bags: a method for biogeochemical process studies. *Marine Ecology Progress Series*, **208**, 273–282.

Hansen L. K., Jakobsen R. and Postma D. (2001). Methanogenesis in a shallow sandy aquifer, Rømø, Denmark. *Geochimica et Cosmochimica Acta*, **65**(17), 2925–2935.

Harbaugh A. W., Banta E. R., Hill M. C. and McDonald M. G. (2000). MODFLOW-2000, the U.S. Geological Survey Modular Ground-Water Model – User guide to Modularization Concepts and the Ground-Water Flow Process. U.S. Geological Survey, Reston, Virginia, p. 121.

Hartog N., Griffioen J. and van der Weijden C. H. (2002). Distribution and reactivity of O_2-reducing components in sediments from a layered aquifer. *Environmental Science & Technology*, **36**(11), 2338–2344.

Hartog N., van Bergen P. F., de Leeuw J. W. and Griffioen J. (2004). Reactivity of organic matter in aquifer sediments: geological and geochemical controls. *Geochimica et Cosmochimica Acta*, **68**(6), 1281–1292.

Hecht-Mendez J., Molina-Giraldo N., Blum P. and Bayer P. (2010). Evaluating MT3DMS for heat transport simulation of closed geothermal systems. *Ground Water*, **48**(5), 741–756.

Helgeson H. C. (1971). Kinetics of mass transfer among silicates and aqueous solutions. *Geochimica et Cosmochimica Acta*, **35**(5), 421–469.

Her N., Amy G., McKnight D., Sohn J. and Yoon Y. (2003). Characterization of DOM as a function of MW by fluorescence EEM and HPLC-SEC using UVA, DOC, and fluorescence detection. *Water Research*, **37**(17), 4295–4303.

Hering J. G., Chen P. Y., Wilkie J. A. and Elimelech M. (1997). Arsenic removal from drinking water during coagulation. *Journal of Environmental Engineering*, **123**(8), 800–807.

Hijnen W. A. M. and Van der Kooij D. (1992). The effect of low concentrations of Assimilable Organic Carbon (AOC) in water on biological clogging of sand beds. *Water Research*, **26**(7), 963–972.

Hijnen W. A. M., Brouwer-Hanzens A. J., Charles K. J. and Medema G. J. (2005). Transport of MS2 phage, *Escherichia coli*, *Clostridium perfringens*, *Cryptosporidium parvum*, and *Giardia intestinalis* in a gravel and a sandy soil. *Environmental Science & Technology*, **39**(20), 7860–7868.

Hoehler T. M., Alperin M. J., Albert D. B. and Martens C. S. (1998). Thermodynamic control on hydrogen concentrations in anoxic sediments. *Geochimica et Cosmochimica Acta*, **62**(10), 1745–1756.

Holm T. R., Eisenreich S. J., Rosenberg H. L. and Holm N. P. (1987). Groundwater geochemistry of short-term aquifer thermal energy storage test cycles. *Water Resources Research*, **23**(6), 1005–1019.

Hong H.-J., Yang J.-S., Kim B.-K. and Yang J.-W. (2011). Arsenic removal behavior by Fe-Al binary oxide: thermodynamic and kinetic study. *Separation Science and Technology*, **46**, 2531–2538.

Houben G. and Trestakis C. (2007). Water Well Rehabilitation and Reconstruction. McGraw Hill, New York.

Hoyer M., Hallgren J., Eisenreich S. and Sterling R. (1994). Field-test results of aquifer thermal energy storage at St. Paul, Minnesota. *Journal of Energy Engineering-Asce*, **120**(2), 67–85.

Huang Y.-H., Hsueh C.-L., Huang C.-P., Su L.-C. and Chen C.-Y. (2007). Adsorption thermodynamic and kinetic studies of Pb(II) removal from water onto a versatile Al_2O_3-supported iron oxide. *Separation and Purification Technology*, **55**(1), 23–29.

Hubert C., Loy A., Nickel M., Arnosti C., Baranyi C., Brüchert V., Ferdelman T., Finster K., Christensen F. M., De Rezende J. R., Vandieken V. and Jørgensen B. B. (2009). A Constant Flux of diverse thermophilic bacteria into the cold arctic seabed. *Science*, **325**(5947), 1541–1544.

Hubert C., Arnosti C., Brüchert V., Loy A., Vandieken V. and Jørgensen B. B. (2010). Thermophilic anaerobes in Arctic marine sediments induced to mineralize complex organic matter at high temperature. *Environmental Microbiology*, **12**(4), 1089–1104.

IF Technology (2007). Common practice in ATES application, field data of 67 projects. Unpublished report. p. 31, Arnhem. http://www.sikb.nl/upload/documents/WKO/Rapport%20energiebesparing%20kwo%20in%20de%20praktijk.pdf (accessed 1 July 2013)

Illieva D., Morasch B. and Haderlein S. B. (2012). *Risikominimierung beim Einsatz von Additiven in Wärmeträgerflüssigkeiten: Risikominimierung beim Einsatz von Additiven in Wärmeträgerflüssigkeiten, Eberhard-Karls Universität Tübingen* (Risk reduction in the use of additives in heat transfer fluids: Influence of Additives on the environmental fate and degradation in the subsurface). http://www.fachdokumente.lubw.baden-wuerttemberg.de/servlet/is/105598/zo4e28006_haderlein.pdf?command=downloadContent&filename=zo4e28006_haderlein.pdf&FIS=203 (accessed 1 July 2013)

Isaksen M. F., Bak F. and Jørgensen B. B. (1994). Thermophilic sulfate-reducing bacteria in cold marine sediment. *Fems Microbiology Ecology*, **14**(1), 1–8.

Jacobson J., Frenz J. and Horváth C. (1984). Measurement of adsorption isotherms by liquid chromatography. *Journal of Chromatography A*, **316**(0), 53–68.

Jakobsen R. and Cold L. (2007). Geochemistry at the sulfate reduction-methanogenesis transition zone in an anoxic aquifer–a partial equilibrium interpretation using 2D reactive transport modeling. *Geochimica et Cosmochimica Acta*, **71**(8), 1949–1966.

Jakobsen R. and Postma D. (1994). In situ rates of sulfate reduction in an aquifer (Rømø, Denmark) and implications for the reactivity of organic matter. *Geology*, **22**(12), 1101–1106.

Jakobsen R. and Postma D. (1999). Redox zoning, rates of sulfate reduction and interactions with Fe-reduction and methanogenesis in a shallow sandy aquifer, Rømø, Denmark. *Geochimica et Cosmochimica Acta*, **63**(1), 137–151.

Jakobsen R., Albrechtsen H.-J., Rasmussen M., Bay H., Bjerg P. L. and Christensen T. H. (1998). H_2 concentrations in a Landfill Leachate Plume (Grindsted, Denmark): *in situ* energetics of terminal electron acceptor processes. *Environmental Science & Technology*, **32**(14), 2142–2148.

James R. O. and Healy T. W. (1972). Adsorption of hydrolyzable metal ions at the oxide – water interface. III. A thermodynamic model of adsorption. *Journal of Colloid and Interface Science*, **40**(1), 65–81.

Jardine P. M. and Sparks D. L. (1984). Potassium-calcium exchange in a multireactive soil system: I. kinetics. *Soil Sci. Soc. Am. J.*, **48**(1), 39–45.

Jenne E. A., Andersson O. and Willemsen A. (1992). Well, hydrology, and geochemistry problems encountered in ATES systems and their solutions. In: Proceedings of the Intersociety Energy Conversion Engineering Conference, San Diego, CA, USA, D. W. Allen (ed.), IEEE, San Diego, CA, USA, pp. 77–88.

Jessen S., Postma D., Larsen F., Nhan P. Q., Hoa L. Q., Trang P. T. K., Long T. V., Viet P. H. and Jakobsen R. (2012). Surface complexation modeling of groundwater arsenic mobility: results of a forced gradient experiment in a Red River flood plain aquifer, Vietnam. *Geochimica et Cosmochimica Acta*, **98**, 186–201.

Jesußek A., Grandel S. and Dahmke A. (2012). Impacts of subsurface heat storage on aquifer hydrogeochemistry. *Environmental Earth Sciences*, **69**(6), 1–14.

Johnson B. B. (1990). Effect of pH, temperature, and concentration on the adsorption of cadmium on goethite. *Environmental Science & Technology*, **24**(1), 112–118.

Jørgensen B. B., Zawacki L. X. and Jannasch H. W. (1990). Thermophilic bacterial sulfate reduction in deep-sea sediments at the Guaymas Basin hydrothermal vent site (Gulf of California). Deep sea research part A. *Oceanographic Research Papers*, **37**(4), 695–710.

Kallmeyer J. and Boetius A. (2004). Effects of Temperature and pressure on sulfate reduction and anaerobic oxidation of methane in hydrothermal sediments of Guaymas basin. *Applied and Environmental Microbiology*, **70**(2), 1231–1233.

Kanel S. R., Manning B., Charlet L. and Choi H. (2005). Removal of Arsenic(III) from groundwater by nanoscale zero-valent iron. *Environmental Science & Technology*, **39**(5), 1291–1298.

Karlsen R. H., Smits F. J. C., Stuyfzand P. J., Olsthoorn T. N. and van Breukelen B. M. (2012). A post audit and inverse modeling in reactive transport: 50 years of artificial recharge in the Amsterdam Water Supply Dunes. *Journal of Hydrology*, **454–455**(0), 7–25.

Kersten M. and Vlasova N. (2009a). Arsenite adsorption on goethite at elevated temperatures. *Applied Geochemistry*, **24**(1), 32–43.

Kersten M. and Vlasova N. (2009b). Silicate adsorption by goethite at elevated temperatures. *Chemical Geology*, **262**(3–4), 336–343.

Kjeldsen P. (1993). Evaluation of gas diffusion through plastic materials used in experimental and sampling equipment. *Water Research*, **27**(1), 121–131.

Klotz D., Seiler K. P., Moser H. and Neumaier F. (1980). Dispersivity and velocity relationship from laboratory and field experiments. *Journal of Hydrology*, **45**(3–4), 169–184.

Klotzbucher T., Kappler A., Straub K. L. and Haderlein S. B. (2007). Biodegradability and groundwater pollutant potential of organic anti-freeze liquids used in borehole heat exchangers. *Geothermics*, **36**(4), 348–361.

Knappett P. S. K., McKay L. D., Layton A., Williams D. E., Alam M. J., Mailloux B. J., Ferguson A. S., Culligan P. J., Serre M. L., Emch M., Ahmed K. M., Sayler G. S. and Geen A. V. (2012). Unsealed tubewells lead to increased fecal contamination of drinking water. *Journal of Water and Health*, **10**(4), 565–578.

Knauss K. G. and Copenhaver S. A. (1995). The solubility of p-xylene in water as a function of temperature and pressure and calculated thermodynamic quantities. *Geochimica et Cosmochimica Acta*, **59**(12), 2443–2448.

Knauss K. G., Dibley M. J., Leif R. N., Mew D. A. and Aines R. D. (2000). The aqueous solubility of trichloroethene (TCE) and tetrachloroethene (PCE) as a function of temperature. *Applied Geochemistry*, **15**(4), 501–512.

Kooi H. (2008). Spatial variability in subsurface warming over the last three decades; insight from repeated borehole temperature measurements in The Netherlands. *Earth and Planetary Science Letters*, **270**(1–2), 86–94.

Korom S. F. (1992). Natural denitrification in the saturated zone: a review. *Water Resources Research*, **28**(6), 1657–1668.

Kraneveld E. A., Buijs M. J., Bonder M. J., Visser M., Keijser B. J. F., Crielaard W. and Zaura E. (2012). The relation between oral candida load and bacterial microbiome profiles in Dutch older adults. *PLoS ONE*, **7**(8), e42770.

Lacombe S., Sudicky E. A., Frape S. K. and Unger A. J. A. (1995). Influence of leaky boreholes on cross-formational groundwater flow and contaminant transport. *Water Resources Research*, **31**(8), 1871–1882.

Lam S. Y., Yeung R. C. Y., Yu T.-H., Sze K.-H. and Wong K.-B. (2011). A rigidifying salt-bridge favors the activity of thermophilic enzyme at high temperatures at the expense of low-temperature activity. *PLoS Biol*, **9**(3), e1001027.

Langwaldt J. H. and Puhakka J. A. (2000). On-site biological remediation of contaminated groundwater: a review. *Environmental Pollution*, **107**(2), 187–197.

LaRowe D. E. and Van Cappellen P. (2011). Degradation of natural organic matter: a thermodynamic analysis. *Geochimica et Cosmochimica Acta* **75**(8), 2030–2042.

LaRowe D. E., Dale A. W., Amend J. P. and Van Cappellen P. (2012). Thermodynamic limitations on microbially catalyzed reaction rates. *Geochimica et Cosmochimica Acta*, **90**, 96–109.

Lerm S., Alawi M., Miethling-Graff R., Wolfgramm M., Rauppach K., Seibt A. and Würdemann H. (2011). Influence of microbial processes on the operation of a cold store in a shallow aquifer: impact on well injectivity and filter lifetime. *Grundwasser*, June, **16**(2), 93–104.

Lerm S., Westphal A., Miethling-Graff R., Alawi M., Seibt A., Wolfgramm M. and Würdemann H. (2013). Thermal effects on microbial composition and microbiologically induced corrosion and mineral precipitation affecting operation of a geothermal plant in a deep saline aquifer. *Extremophiles*, **17**(2), 311–327.

Li W., Pan G., Zhang M., Zhao D., Yang Y., Chen H. and He G. (2008). EXAFS studies on adsorption irreversibility of Zn(II) on TiO2: temperature dependence. *Journal of Colloid and Interface Science*, **319**(2), 385–391.

Limousin G., Gaudet J. P., Charlet L., Szenknect S., Barthès V. and Krimissa M. (2007). Sorption isotherms: a review on physical bases, modeling and measurement. *Applied Geochemistry*, **22**(2), 249–275.

Lovley D. R. and Coates J. D. (2000). Novel forms of anaerobic respiration of environmental relevance. *Current Opinion in Microbiology*, **3**(3), 252–256.

Lund J. W. and Bertani R. (2010). Worldwide geothermal utilization 2010, Geothermal Resources Council Annual Meeting 2010, Sacramento, 24–27 October 2010; Sacramento, pp. 182–185.

Mahmood T., Saddique M. T., Naeem A., Mustafa S., Zeb N., Shah K. H. and Waseem M. (2011). Kinetic and thermodynamic study of Cd(II), Co(II) and Zn(II) adsorption from aqueous solution by NiO. *Chemical Engineering Journal*, **171**(3), 935–940.

Maire P., Blunier P., Parriaux A. and Tacher L. (2006). Underground planning and optimization of the underground resources combination looking for sustainable development in urban areas. Going Underground: Excavating the Subterranean City, Manchester, UK, 21–22 September 2006, The Geological Society of London, Manchester, UK.

Maji S. K., Pal A., Pal T. and Adak A. (2007). Adsorption thermodynamics of arsenic on laterite soil. pp. 161–176.

Manju G. N., Anoop Krishnan K., Vinod V. P. and Anirudhan T. S. (2002). An investigation into the sorption of heavy metals from wastewaters by polyacrylamide-grafted iron(III) oxide. *Journal of Hazardous Materials*, **91**(1–3), 221–238.

Marvin-DiPasquale M. C. and Capone, D. G. (1998). Benthic sulfate reduction along the Chesapeake Bay central channel. I. Spatial trends and controls. *Marine Ecology Progress Series*, **168**, 213–228.

Mayo A. (2010). Ambient well-bore mixing, aquifer cross-contamination, pumping stress, and water quality from long-screened wells: what is sampled and what is not? *Hydrogeology Journal,* June, **18**(4), 823–837.

McBee R. H. and McBee V. H. (1956). The incidence of thermophilic bacteria in arctic soils and waters. *Journal of Bacteriology*, **71**(2), 182–185.

Mendizabal I., Baggelaar P. K. and Stuyfzand P. J. (2012). Hydrochemical trends for public supply well fields in The Netherlands (1898–2008), natural backgrounds and upscaling to groundwater bodies. *Journal of Hydrology*, **450–451**(0), 279–292.

Meyer C. F. and Todd D. K. (1973). Conserving energy with heat storage wells. *Environmental Science & Technology*, **7**(6), 512–516.

Middelburg J. J. (1989). A simple rate model for organic matter decomposition in marine sediments. *Geochimica et Cosmochimica Acta*, **53**(7), 1577–1581.

Mirecki J. E., Bennett M. W. and López-Baláez M. C. (2012). Arsenic control during aquifer storage recovery cycle tests in the floridan aquifer. *Ground Water*, **51**(4), 539–549.

Mustafa G., Kookana R. S. and Singh B. (2006). Desorption of cadmium from goethite: effects of pH, temperature and aging. *Chemosphere*, **64**(5), 856–865.

Mustafa Omer A. (2008). Ground-source heat pumps systems and applications. *Renewable and Sustainable Energy Reviews*, **12**(2), 344–371.

Mustafa S., Waseem M., Naeem A., Shah K. H., Ahmad T. and Safdar M. (2010). Effect of temperature on Cd2+ sorption by mixed oxides of iron and silicon. *Chinese Journal of Chemistry*, **28**(11), 2204–2208.

Muyzer G. and Stams A. J. M. (2008). The ecology and biotechnology of sulphate-reducing bacteria. *Nature Reviews Microbiology*, **6**(6), 441–454.

Negrea A., Lupa L., Ciopec M., Lazau R., Muntean C. and Negrea P. (2010). Adsorption of As(III) ions onto iron-containing waste sludge. *Adsorption Science & Technology*, **28**(6), 467–484.

Ni Z., Grotenhuis T., Hartog N. and Rijnaarts H. (2011). Effect of Aquifer Thermal Energy Storage (ATES) on Bioremediation of Chlorinated Solvents, 1ste Nationaal Congres Bodemenergie, Utrecht, 13–14 October 2011, Utrecht.

Ni Z., Smit M., Grotenhuis T., Hartog N. and Rijnaarts H. (2013). Findings on Limiting Factors of Intrinsic Bioremediation of Chlorinated Ethenes Combined with Aquifer Thermal Energy Storage (ATES), Aquaconsoil, Barcalona, 16–19 April 2013; Barcalona.

Noyes P. D., McElwee M. K., Miller H. D., Clark B. W., Van Tiem L. A., Walcott K. C., Erwin K. N. and Levin E. D. (2009). The toxicology of climate change: environmental contaminants in a warming world. *Environment International*, **35**(6), 971–986.

Oostrum N., Stuurman R., Oude Essink G. and Aguilera D. R. (2008). Evaluatie monitoringsvoorschriften wko van de provincie Flevoland, Deltares, Utrecht, p. 102.

Overmann J. (2013). Principles of enrichment, isolation, cultivation, and preservation of prokaryotes. In: The Prokaryotes, E. Rosenberg, E. DeLong, S. Lory, E. Stackebrandt and F. Thompson (eds), Springer, Berlin Heidelberg, pp. 149–207.

Ovreås L., Forney L., Daae F. L. and Torsvik V. (1997). Distribution of bacterioplankton in meromictic Lake Saelenvannet, as determined by denaturing gradient gel electrophoresis of PCR-amplified gene fragments coding for 16S rRNA. *Applied and Environmental Microbiology*, **63**(9), 3367–3373.

Pallud C. and Van Cappellen P. (2006). Kinetics of microbial sulfate reduction in estuarine sediments. *Geochimica et Cosmochimica Acta*, **70**(5), 1148–1162.

Pallud C., Meile C., Laverman A. M., Abell J. and Van Cappellen P. (2007). The use of flow-through sediment reactors in biogeochemical kinetics: methodology and examples of applications. *Marine Chemistry*, **106**(1–2), 256–271.

Palmer C. D. and Cherry J. A. (1984). Geochemical reactions associated with low-temperature thermal energy storage in aquifers. *Can. Geotech. J.*, **21**(3), 13.

Parkhurst D. L. and Appelo C. A. J. (1999). User's guide to PHREEQC (version 2) – a computer program for speciation, batch-reaction, one-dimensional transport, and inverse geochemical calculations, U.S. Geological Survey, p. 312.

Perfect E., Sukop M. C. and Haszler G. R. (2002). Prediction of dispersivity for undisturbed soil columns from water retention parameters. *Soil Sci. Soc. Am. J.*, **66**(3), 696–701.

Perlinger J. A., Almendinger J. E., Urban N. R. and Eisenreich S. J. (1987). Groundwater geochemistry of aquifer thermal energy storage: long-term test cycle. *Water Resources Research*, **23**(12), 2215–2226.

Philippacopoulos A. J. and Berndt M. L. (2001). Influence of debonding in ground heat exchangers used with geothermal heat pumps. *Geothermics*, **30**(5), 527–545.

Postma D., Larsen F., Minh Hue N. T., Duc M. T., Viet P. H., Nhan P. Q. and Jessen S. (2007). Arsenic in groundwater of the Red River floodplain, Vietnam: controlling geochemical processes and reactive transport modeling. *Geochimica et Cosmochimica Acta*, **71**(21), 5054–5071.

Prommer H. and Stuyfzand P. J. (2005). Identification of temperature-dependent water quality changes during a deep well injection experiment in a pyritic aquifer. *Environmental Science & Technology*, **39**(7), 2200–2209.

Pruesse E., Quast C., Knittel K., Fuchs B. M., Ludwig W., Peplies J. and Glöckner F. O. (2007). SILVA: a comprehensive online resource for quality checked and aligned ribosomal RNA sequence data compatible with ARB. *Nucleic Acids Research*, **35**(21), 7188–7196.

Ramamoorthy M., Jin H., Chiasson A. D. and Spitler J. D. (2001). Optimal sizing of hybrid ground-source heat pump systems that use a cooling pond as a supplemental heat rejecter – a system simulation approach. Atlanta, GA, United States, pp. 26–38.

Reeder J. and Knight R. (2010). Rapidly denoising pyrosequencing amplicon reads by exploiting rank-abundance distributions. *Nature Methods*, **7**(9), 668–669.

Reemtsma T., Bredow A. and Gehring M. (1999). The nature and kinetics of organic matter release from soil by salt solutions. *European Journal of Soil Science*, **50**(1), 53–64.

Reichert P. and Omlin M. (1997). On the usefulness of overparameterized ecological models. *Ecological Modelling*, **95**(2–3), 289–299.

Reuss M. and Konstantinidou E. (2006). 10 Years VDI 4640 – German guidelines for ground coupled heat pumps, UTES and direct thermal use of the underground. Proc. Ecostock 2006, Richard Stockton College New Jersey, Richard Stockton College New Jersey.

Riemsdijk W. H. v. and Hiemstra T. (2006). Chapter 8 The CD-MUSIC model as a framework for interpreting ion adsorption on metal (hydr) oxide surfaces. In: Interface Science and Technology, L. Johannes, (ed.), Vol. 11, Elsevier, pp. 251–268.

Rittmann B. E., Stilwell D., Garside J. C., Amy G. L., Spangenberg C., Kalinsky A. and Akiyoshi E. (2002). Treatment of a colored groundwater by ozone-biofiltration: pilot studies and modeling interpretation. *Water Research*, **36**(13), 3387–3397.

Robador A., Brüchert V. and Jørgensen B. B. (2009). The impact of temperature change on the activity and community composition of sulfate-reducing bacteria in arctic versus temperate marine sediments. *Environmental Microbiology*, **11**(7), 1692–1703.

Robador A., Bruchert V., Steen A. D. and Arnosti C. (2011). Temperature induced decoupling of enzymatic hydrolysis and carbon remineralization in long-term incubations of Arctic and temperate sediments. *Geochimica et Cosmochimica Acta*, **74**(8), 2316–2326.

Rock G. and Kupfersberger H. (2002). Numerical delineation of transient capture zones. *Journal of Hydrology*, **269**(3–4), 134–149.

Rodda D. P., Johnson B. B. and Wells J. D. (1993). The Effect of Temperature and pH on the Adsorption of Copper(II), Lead(II), and Zinc(II) onto Goethite. *Journal of Colloid and Interface Science*, **161**(1), 57–62.

Roychoudhury A. N. (2004). Sulfate respiration in extreme environments: a kinetic study. *Geomicrobiology Journal*, **21**(1), 33–43.

Roychoudhury A. N., Viollier E. and Van Cappellen P. (1998). A plug flow-through reactor for studying biogeochemical reactions in undisturbed aquatic sediments. *Applied Geochemistry*, **13**(2), 269–280.

Sanner B. (2007). Guidelines, standards, certification and legal permits for ground source heat pumps in the european union, 9th International IEA Heat Pump Conference, 20–22 May 2008, Zürich, Switzerland, Zürich, Switzerland.

Sanner B., Karytsas C., Mendrinos D. and Rybach L. (2003). Current status of ground source heat pumps and underground thermal energy storage in Europe. *Geothermics*, **32**(4–6), 579–588.

Santi P. M., McCray J. E. and Martens J. L. (2006). Investigating cross-contamination of aquifers. *Hydrogeology Journal*, **14**(1–2), 51–68.

Sawhney B. L. (1966). Kinetics of cesium sorption by clay mineralsl. *Soil Sci. Soc. Am. J.*, **30**(5), 565–569.

Sawicka J. E., Jørgensen B. B. and Brüchert V. (2012). Temperature characteristics of bacterial sulfate reduction in continental shelf and slope sediments. *Biogeosciences Discuss.*, **9**(1), 673–700.

Schippers A. and Reichling J. (2006). Laboruntersuchungen zum Einfluss von Temperaturveränderungen auf die Mikrobiologie des Untergrundes. *Grundwasser*, **11**(1), 40–45.

Schmidt K. R., Conrad R., Hollert H. and Tiehm A. (2013). Risk assessment of geothermal heat transfer fluids, Aquaconsoil, Barcalona, 16–19 April 2013, Barcalona.

Sharma L., Greskowiak J., Ray C., Eckert P. and Prommer H. (2012). Elucidating temperature effects on seasonal variations of biogeochemical turnover rates during riverbank filtration. *Journal of Hydrology*, **428–429**, 104–115.

Shipley H. J., Engates K. E. and Grover V. A. (2013). Removal of Pb(II), Cd(II), Cu(II), and Zn(II) by hematite nanoparticles: effect of sorbent concentration, pH, temperature, and exhaustion. *Environmental Science and Pollution Research*, **20**(3), 1727–1736.

Slenders H. L. A., Dols P., Verburg R. and de Vries A. J. (2010). Sustainable remediation panel: sustainable synergies for the subsurface: combining groundwater energy with remediation. *Remediation Journal*, **20**(2), 143–153.

Smedley P. L. and Kinniburgh D. G. (2002). A review of the source, behaviour and distribution of arsenic in natural waters. *Applied Geochemistry*, **17**(5), 517–568.

Smeets P. W. M. H., Medema G. J. and van Dijk J. C. (2009). The Dutch secret: how to provide safe drinking water without chlorine in the Netherlands. *Drinking Water Engineering and Science*, **2**(1), 1–14.

Sowers L., York K. P. and Stiles L. (2006). Impact of Thermal Buildup on Groundwater Chemistry and Aquifer Microbes, Ecostock 2006, New Jersey, New Jersey.

Sparks D. L. and Jardine P. M. (1984). Comparison of kinetic equations to describe potassium-calcium exchange in pure and in mixed systems. *Soil Science*, **138**(2), 115–122.

Sposito G. (2008). The Chemistry of Soils, 2nd edn. Oxford University Press, New York, p. 344.

Stollenwerk K. G., Breit G. N., Welch A. H., Yount J. C., Whitney J. W., Foster A. L., Uddin M. N., Majumder R. K. and Ahmed N. (2007). Arsenic attenuation by oxidized aquifer sediments in Bangladesh. *Science of The Total Environment*, **379**(2–3), 133–150.

Stumm W. and Morgan J. (1996). Aquatic Chemistry: Chemical Equilibria and Rates in Natural Waters, 3rd edn. John Wiley & Sons, New York, p. 1040.

Stuyfzand P. J. (1998). Quality changes upon injection into anoxic aquifers in the Netherlands: evaluation of 11 experiments. In: 3rd Intern. Symp. on Artificial Recharge, Amsterdam, J. Peters, (ed.), Balkema, Amsterdam, pp. 283–291.

Stuyfzand P. J. (1999). Patterns in groundwater chemistry resulting from groundwater flow. *Hydrogeology Journal*, **7**(1), 15–27.

Stuyfzand P. J. (2001). Pyrite oxidation and side-reactions upon deep well injection. In 10th International Symposium on Water Rock Interaction, Villasimius, Italy, 10–15 June 2001, Villasimius, Italy, pp. 1151–1154.

Stuyfzand P. J., Lebbink J. and Nienhuis P. (2008). Aquifer Thermal Energy Storage in Groundwater Protection Areas: (possible) Objections (in Dutch). KWR Watercycle Research Institute, Nieuwegein, p. 32

Stuyfzand P. J., Rossum P. v. and Mendizabal I. (2006). Does arsenic, in groundwater of the compound Rhine-Meuse-Scheldt-Ems delta, mencase drinking water supply in the Netherlands? In: Arsenic in Groundwater, a World Problem, Utrecht, C. A. J. Appelo. (ed.), Unesco, WMO, Deltares & IGRAC, Utrecht, pp. 102–125.

Sverdrup H. U. (1990). The Kinetics of Base Cation Release Due to Chemical Weathering. Lund University Press, Lund.

Sverjensky D. A. and Sahai N. (1998). Theoretical prediction of single-site enthalpies of surface protonation for oxides and silicates in water. *Geochimica et Cosmochimica Acta*, **62**(23–24), 3703–3716.

Swedlund P. J. and Webster J. G. (1999). Adsorption and polymerisation of silicic acid on ferrihydrite, and its effect on arsenic adsorption. *Water Research*, **33**(16), 3413–3422.

Taniguchi M. and Uemura T. (2005). Effects of urbanization and groundwater flow on the subsurface temperature in Osaka, Japan. *Physics of The Earth and Planetary Interiors*, **152**(4), 305–313.

Tarpgaard I. H., Røy H. and Jørgensen B. B. (2011). Concurrent low- and high-affinity sulfate reduction kinetics in marine sediment. *Geochimica et Cosmochimica Acta*, **75**(11), 2997–3010.

TCB (2009). Advies duurzaam gebruik van de bodem voor WKO. p. 47. http://www.tcbodem.nl/files/a050%20advies%20duurzaam%20gebruik%20bodem%20voor%20wko.pdf (accessed 1 October 2013)

Ten Hulscher T. E. M. and Cornelissen G. (1996). Effect of temperature on sorption equilibrium and sorption kinetics of organic micropollutants – a review. *Chemosphere* **32**(4), 609–626.

Toride N., Leij F. J. and van Genuchten M. T. (1995). TheCXTFIT Code for Estimating Transport Parameters from Laboratory or Field Tracer Experiments Version 2.0. U. S. Salinity Laboratory, Riverside, California, p. 131.

Trivedi P. and Axe L. (2000). Modeling Cd and Zn sorption to hydrous metal oxides. *Environmental Science & Technology*, **34**(11), 2215–2223.

Trivedi P. and Axe L. (2001). Predicting divalent metal sorption to hydrous Al, Fe, and Mn oxides. *Environmental Science & Technology*, **35**(9), 1779–1784.

Trivedi P., Dyer J. A., Sparks D. L. and Pandya K. (2004). Mechanistic and thermodynamic interpretations of zinc sorption onto ferrihydrite. *Journal of Colloid and Interface Science*, **270**(1), 77–85.

Underground Energy Taskforce (2009). Green Light for Underground Thermal Energy Storage (in Dutch). Ministry of Housing, Spatial Planning and the Environment, Den Haag. http://www.rijksoverheid.nl/documenten-en-publicaties/brochures/2009/03/01/groen-licht-voor-bodemenergie-advies-taskforce-wko.html (accessed 1 January 2013)

Ursej S. and Kontic B. (2007). The role of surface characteristics in directing subsurface spatial planning processes: the case study of a high-speed railway in Slovenia. *Tunnelling and Underground Space Technology*, **22**(4), 414–432.

van Balen R. T., Houtgast R. F. and Cloetingh S. A. P. L. (2005). Neotectonics of The Netherlands: a review. *Quaternary Science Reviews*, **24**(3–4), 439–454.

van Breukelen B. M., Röling W. F. M., Groen J., Griffioen J. and van Verseveld H. W. (2003). Biogeochemistry and isotope geochemistry of a landfill leachate plume. *Journal of Contaminant Hydrology*, **65**(3–4), 245–268.

van Halem D., Moed D. H., Verberk J. Q. J. C., Amy G. L. and van Dijk J. C. (2012). Cation exchange during subsurface iron removal. *Water Research*, **46**(2), 307–315.

van Helvoort P.-J., Griffioen J. and Hartog N. (2007). Characterization of the reactivity of riverine heterogeneous sediments using a facies-based approach; the Rhine-Meuse delta (The Netherlands). *Applied Geochemistry*, **22**(12), 2735–2757.

Vanhoudt D., Desmedt J., Van Bael J., Robeyn N. and Hoes H. (2011). An aquifer thermal storage system in a Belgian hospital: long-term experimental evaluation of energy and cost savings. *Energy and Buildings*, **43**(12), 3657–3665.

Vecchia A. V. and Cooley R. L. (1987). Simultaneous confidence and prediction intervals for nonlinear regression models with application to a groundwater flow model. *Water Resources Research*, **23**(7), 1237–1250.

Vetriani C., Jannasch H. W., MacGregor B. J., Stahl D. A. and Reysenbach A.-L. (1999). Population structure and phylogenetic characterization of marine benthic archaea in deep-sea sediments. *Applied and Environmental Microbiology*, **65**(10), 4375–4384.

Vetter A., Mangelsdorf K., Wolfgramm M., Rauppach K., Schettler G. and Vieth-Hillebrand A. (2012). Variations in fluid chemistry and membrane phospholipid fatty acid composition of the bacterial community in a cold storage groundwater system during clogging events. *Applied Geochemistry*, **27**(6), 1278–1290.

Vierheilig J., Frick C., Mayer R. E., Kirschner A. K. T., Reischer G. H., Derx J., Mach R. L., Sommer R. and Farnleitner A. H. (2013). Clostridium perfringens is not suitable for the indication of fecal pollution from ruminant wildlife but is associated with excreta from non-herbivorous animals and human sewage. *Applied and Environmental Microbiology*, August, **79**(16), 5089–5092.

Visser A., Broers H. P., van der Grift B. and Bierkens M. F. P. (2007). Demonstrating trend reversal of groundwater quality in relation to time of recharge determined by 3H/3He. *Environmental Pollution*, **148**(3), 797–807.

Von Gunten U. (2003). Ozonation of drinking water: Part I. Oxidation kinetics and product formation. *Water Research*, **37**(7), 1443–1467.

Vollrath S., Behrends T., Koch C. B. and Cappellen P. V. (2013). Effects of temperature on rates and mineral products of microbial Fe(II) oxidation by Leptothrix cholodnii at microaerobic conditions. *Geochimica et Cosmochimica Acta*, **108**(0), 107–124.

Wallage Z. E., Holden J. and McDonald A. T. (2006). Drain blocking: an effective treatment for reducing dissolved organic carbon loss and water discolouration in a drained peatland. *Science of The Total Environment*, **367**(2–3), 811–821.

Wallis I., Prommer H., Pichler T., Post V., Norton S., Annable M. D. and Simmons C. T. (2011). Process-based reactive transport model to quantify arsenic mobility during aquifer storage and recovery of potable water. *Environmental Science & Technology*, **45**(16), 6924–6931.

Wesolowski D. J., Machesky M. L., Ridley M. K., Palmer D. A., Zhang Z., Fenter P. A., Predota M. and Cummings P. T. (2008). Ion adsorption on metal oxide surfaces to hydrothermal conditions. *ECS Transactions*, **11**(27), 167–180.

Westerhoff W. E. (2003). Description of the Lithostratigraphical Unit: Sterksel Formation (in Dutch). TNO-NITG, p. 6. http://www.dinoloket.nl/formatie-van-sterksel (accessed 1 July 2013)

Weston N. B. and Joye S. B. (2005). Temperature-driven decoupling of key phases of organic matter degradation in marine sediments. *Proceedings of the National Academy of Sciences of the United States of America*, **102**(47), 17036–17040.

White A. F. and Brantley S. L. (2003). The effect of time on the weathering of silicate minerals: why do weathering rates differ in the laboratory and field? *Chemical Geology*, **202**(3–4), 479–506.

Wielen P. W. J. J. v. d., Senden W. J. M. K. and Medema G. (2008). Removal of bacteriophages MS2 and X174 during transport in a sandy anoxic aquifer. *Environmental Science & Technology*, **42**(12), 4589–4594.

Wijnja H. and Schulthess C. P. (2002). Effect of carbonate on the adsorption of selenate and sulfate on goethite. *Soil Science Soceity American Journal*, **66**(4), 1190–1197.

Winters A. L. (1992). Summary of Research on Microbiological Processes. Deptartment of Biological Sciences. University of Alabama, p. 64.

Witteveen+Bos (2008). Masterplan UTES Goudse Poort Gouda. Report to the municipality of Gouda. Unpublished Report. Witteveen+Bos, Almere, p. 36. http://www.goudsepoort.nl/files/organisatie/gemeente/masterplan_wko.pdf (accessed 1 August 2013)

Wollast R. (1967). Kinetics of the alteration of K-feldspar in buffered solutions at low temperature. *Geochimica et Cosmochimica Acta*, **31**(4), 635–648.

Wuijts S. and Rijswick H. F. M. W. (2007). Drinkwateraspecten en de Kaderrichtlijn Water. Rijksinstituut voor Volksgezondheid en Milieu, p. 92.

Xu N. and Saiers J. E. (2010). Temperature and hydrologic controls on dissolved organic matter mobilization and transport within a forest topsoil. *Environmental Science & Technology*, **44**(14), 5423–5429.

Zaadnoordijk W. J., Hornstra L. M. and Bonte M. (2013). Afwegingskader hoge temperatuur warmte koude opslag voor de provincie Gelderland (draft). KWR Watercycle research institute, p. 33.

Zhang P. C. and Sparks D. L. (1990). Kinetics and mechanisms of sulfate adsorption/desorption on goethite using pressure-jump relaxation. *Soil Science Soceity American Journal*, **54**(5), 1266–1273.

Zheng C. (1990). MT3D, A Modular Three-dimensional Transport Model for Simulation of Advection, Dispersion and Chemical Reactions of Contaminants in Groundwater Systems. Report to the U.S. Environmental Protection Agency, p. 170.

Zheng C. (2010). MT3DMS v5.3 Supplemental User's Guide. Department of Geological Sciences, The University of Alabama, Tuscaloosa, Alabama, p. 56. www.hydro.geo.ua.edu/mt3d/mt3dms_v5_supplemental.pdf (accessed 1 August 2013)

Zou W., Han R., Chen Z., Jinghua Z. and Shi J. (2006). Kinetic study of adsorption of Cu(II) and Pb(II) from aqueous solutions using manganese oxide coated zeolite in batch mode. *Colloids and Surfaces A: Physicochemical and Engineering Aspects*, **279**(1–3), 238–246.

Zuurbier K., Hartog N., Valstar J. and van Nieuwkerk E. (2010). Sterk verbeterde analyse van interactie warmte-koudeopslag en verontreinigd grondwater. *H2O*, **3**, 33–36.

Zuurbier K., Bakker M., Zaadnoordijk W. and Stuyfzand P. (2013a). Identification of potential sites for aquifer storage and recovery (ASR) in coastal areas using ASR performance estimation methods. *Hydrogeology Journal*, September, **21**(6), 1373–1383.

Zuurbier K. G., Hartog N., Valstar J., Post V. E. A. and van Breukelen B. M. (2013b). The impact of low-temperature seasonal aquifer thermal energy storage (SATES) systems on chlorinated solvent contaminated groundwater: modeling of spreading and degradation. *Journal of Contaminant Hydrology,* **147**(0), 1–13.

Appendix 1

Supporting information

Figure S1.1 Impression of column setup in an 11°C room.

Table S1.1 Complete results of geochemical analysis of sediments A-C used in column experiments.

Method	TGA			Grain size laser		XRF										
Sample	**LOI330**	**LOI550**	**$CaCO_3$tot**	**Median**	**Clay**	**SiO_2**	**Al_2O_3**	**Fe_2O_3**	**CaO**	**MgO**	**Na_2O**	**K_2O**	**MnO**	**TiO_2**	**P_2O_5**	**Cr_2O_3**
	(%)	(%)	(%)	(µm)	(%)	%	%	%	%	%	%	%	%	%	%	%
MDL						0.1	0.01	0.01	0.01	0.01	0.01	0.01	0.01	0.01	0.01	0.001
Sediment H1	0.24	0.76	6.01	187	3.71	82.7	3.59	1.59	3.02	0.86	0.58	1.23	0.04	0.19	0.05	0.037
Sediment H2	0.15	0.48	0.53	244	2.22	89.8	5.41	1.14	0.30	0.32	1.00	1.82	0.02	0.19	0.04	0.028
Sediment S	0.06	0.14	0.73	310	1.34	92.5	3.36	0.59	0.39	0.22	0.52	1.32	0.02	0.11	0.02	0.001

Method	XRF			Leco		Destr. by Lithium metaborate/tetraborate fusion and dilute nitric digestion followed by ICP-MS										
Sample	**Ba**	**LOI**	**SUM**	**C-tot**	**S-tot**	**Ba**	**Be**	**Co**	**Cs**	**Ga**	**Hf**	**Nb**	**Rb**	**Sn**	**Sr**	**Ta**
	%	%	%	%	%	ppm	ppm	ppm	ppm	ppm	ppm	ppm	ppm	ppm	ppm	ppm
MDL	0.01			0.02	0.02	1	1	0.2	0.1	0.5	0.1	0.1	0.1	1	0.5	0.1
Sediment H1	0.02	3.57	100.15	0.86	0.10	231	2	4.2	1.9	3.4	5.6	3.7	41.6	11	93.1	0.4
Sediment H2	0.03	0.71	100.74	0.1	0.03	281	<1	3.7	2	5.3	2.3	3.4	61.7	14	63.3	0.3
Sediment S	0.02	0.89	100.00	0.12	0.05	233	<1	2.6	1.1	2.7	2.7	2.1	42.1	<1	43.6	0.2

Method	Destruction by Lithium metaborate/tetraborate fusion and dilute nitric digestion followed by ICP-MS															
Sample	**Th**	**U**	**V**	**W**	**Zr**	**Y**	**La**	**Ce**	**Pr**	**Nd**	**Sm**	**Eu**	**Gd**	**Tb**	**Dy**	**Ho**
	ppm	ppm	ppm	ppm	ppm	ppm	ppm	ppm	ppm	ppm	ppm	ppm	ppm	ppm	ppm	ppm
MDL	0.2	0.1	8	0.5	0.1	0.1	0.1	0.1	0.02	0.3	0.05	0.02	0.05	0.01	0.05	0.02
Sediment H1	4.3	1.3	18	<0.5	238.3	10.8	13.8	27.8	3.24	12.5	2.34	0.46	2.12	0.31	1.84	0.37
Sediment H2	3.4	0.9	17	<0.5	79.5	10.6	11.6	23.2	2.87	10.4	2.26	0.48	2.07	0.34	1.95	0.38
Sediment S	3.5	1.0	<8	<0.5	101.4	8.4	10.9	21.4	2.46	9.2	1.84	0.37	1.55	0.25	1.50	0.29

Method	As above				Aqua Regia /ICP-MS											
Sample	**Er**	**Tm**	**Yb**	**Lu**	**Mo**	**Cu**	**Pb**	**Zn**	**Ni**	**As**	**Cd**	**Sb**	**Bi**	**Ag**	**Au**	**Hg**
	ppm	ppm	ppm	ppm	ppm	ppm	ppm	ppm	ppm	ppm	ppm	ppm	ppm	ppm	PPB	ppm
MDL	0.03	0.01	0.05	0.01	0.1	0.1	0.1	1	0.1	0.5	0.1	0.1	0.1	0.1	0.5	0.01
Sediment H1	1.08	0.17	1.11	0.18	0.2	11.0	32.1	15	11.1	5.7	<0.1	0.1	<0.1	<0.1	1.8	<0.01
Sediment H2	1.2	0.17	1.13	0.16	5.8	8.7	21.1	19	12.9	2.9	<0.1	0.2	<0.1	<0.1	<0.5	<0.01
Sediment S	0.86	0.12	0.87	0.13	0.1	2.9	3.2	8	7.4	3.7	<0.1	0.1	<0.1	<0.1	1.3	<0.01

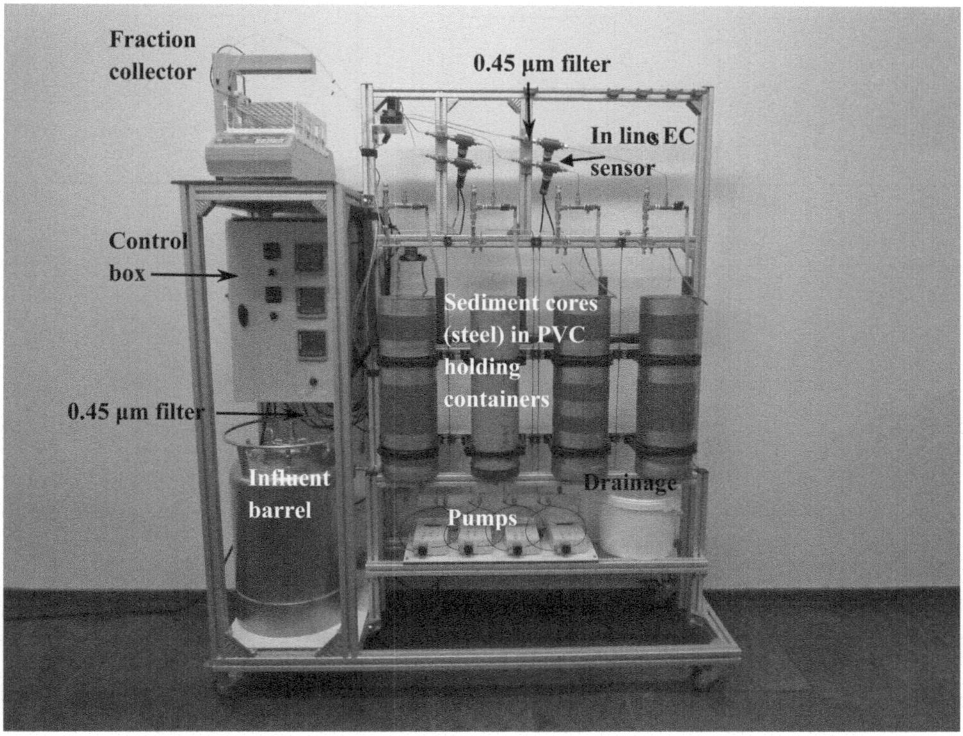

Figure S1.2 Front view of column setup.

Table S1.2 Detection limits for water analyses. All values are in µg L^{-1}.

Element	MDL	Element	MDL	Element	MDL
Na	6	Cd	0.1	S	70
K	4	Co	0.5	Sb	4
Ca	20	Cr	1	Sc	0.03
Mg	10	Cu	0.5	Se	8
Fe	1	Eu	0.1	Sr	0.5
Mn	0.1	Ga	2	Th	2
Al	3	Ho	0.3	Ti	0.2
Si	10	In	4	Tl	4
Ag	0.8	La	40	U	4
As	1	Li	0.3	V	1
B	4	Mo	1	W	2
Ba	0.2	Ni	1	Yb	0.07
Be	0.02	P	6	Zn	0.2
Bi	3	Pb	1	Zr	0.4

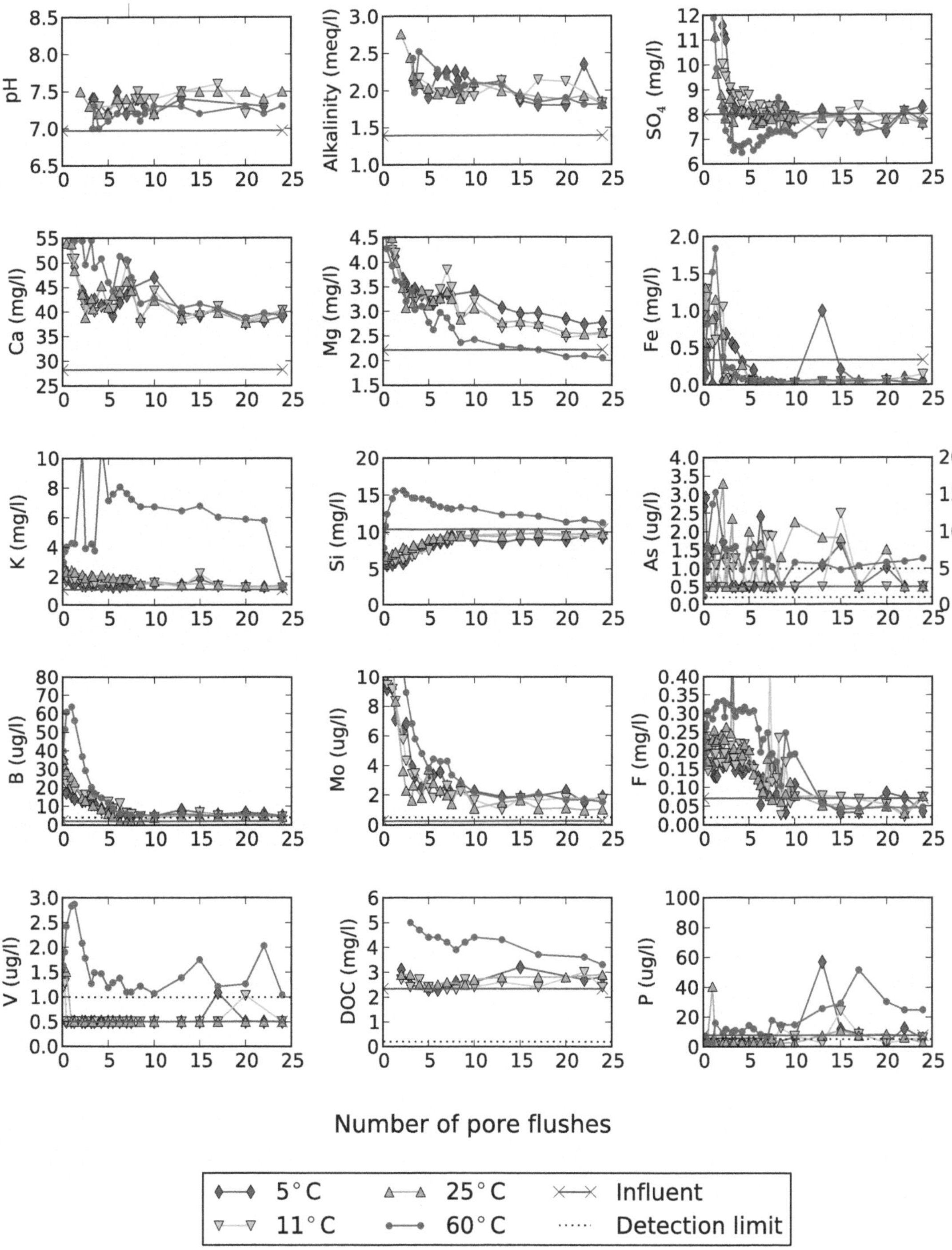

Figure S1.3 Leaching graphs for sediment H1. Note that the 60°C effluent for As is plotted on a secondary y-axis.

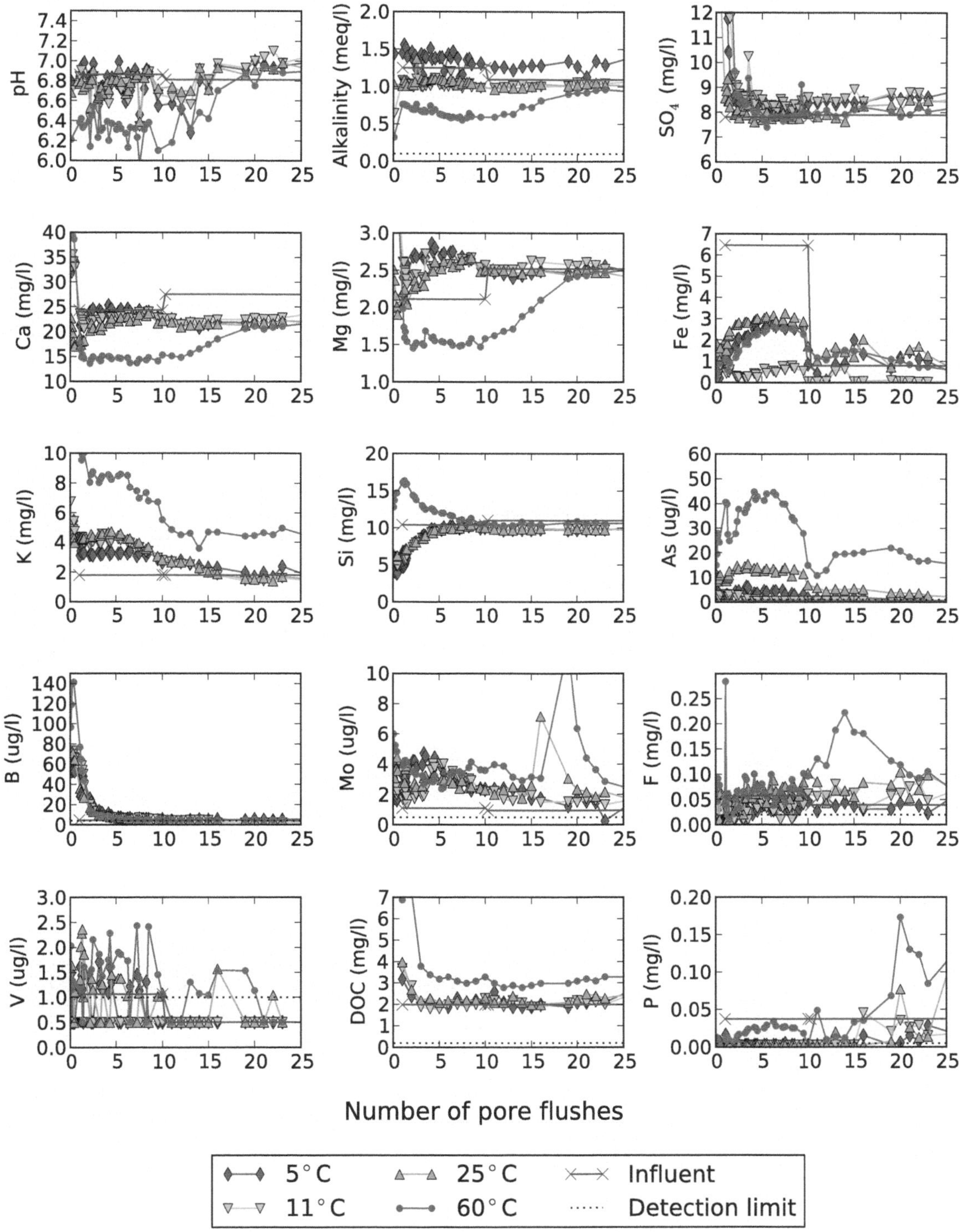

Figure S1.4 Leaching graphs for sediment H2.

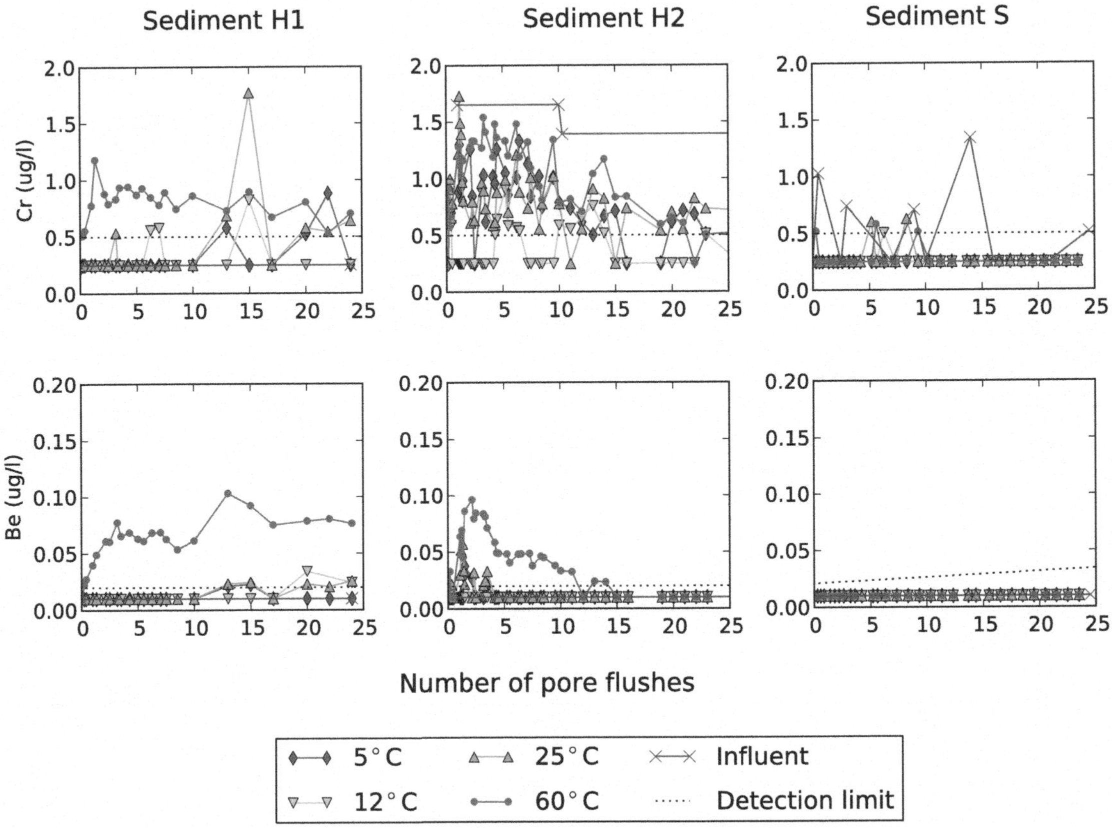

Figure S1.5 Leaching graphs for Cr and Be for all three sediments.

Table S1.3 Results from blank testing at 20°C and 60°C. All values are in μg L^{-1}.

	Demineralized water			Tap water		
	Control	20°C	60°C	Control	20°C	60°C
Fe	<1	75	87	30	6	18
Cr	<1	<1	<1	<1	<1	<1
Ni	<1	7	8	5	3	6
Mo	<1	2	2	3	1	2
V	<1	<1	<1	4	3	2

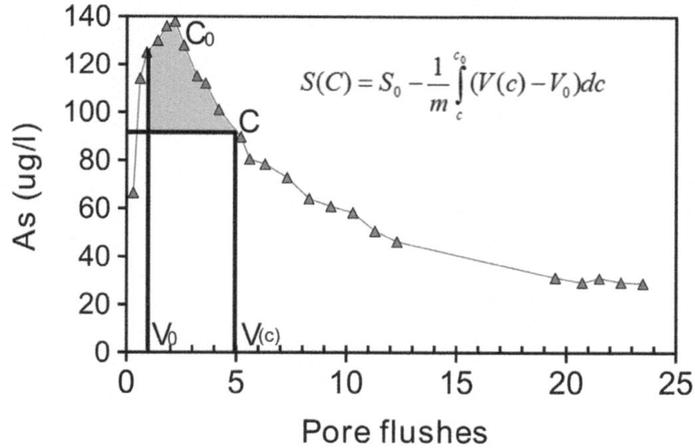

Figure S1.6 Example illustrating how the sorbed amount of arsenic is calculated from the eluent curve.

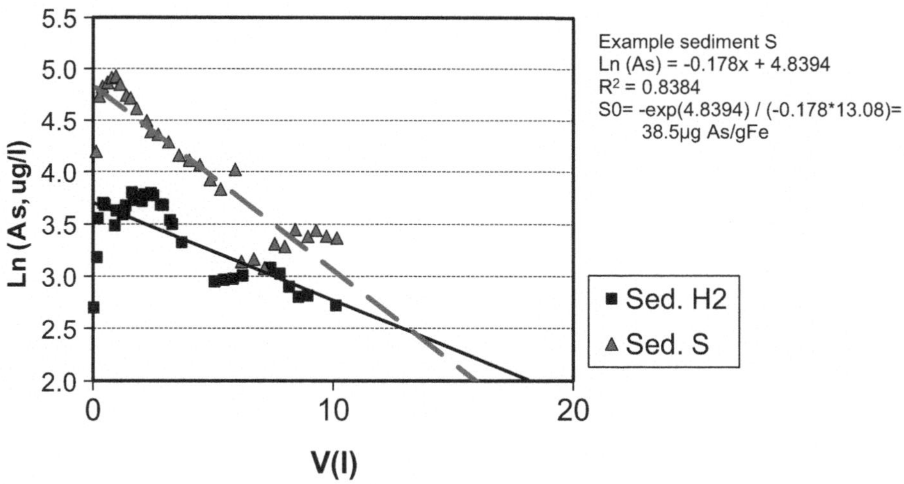

Figure S1.7 Calculation of initial sorbed amount of arsenic (S_0) from a Ln(As) versus leached volume (V) plot.

Appendix 2

Supporting information

S2.1 EXPERIMENTAL DETAILS

The sediment cores were first flushed for 25 days at 0.37 and 0.33 mL min^{-1} for sediments H and S, respectively (resulting in a water residence time of 1 day), allowing the microbiological community to acclimatize to the temperatures and to investigate leaching behavior (Chapter 4). Following the 25 days, the residence time of water in the sediments was stepwise increased resulting in residence times of 2, 10, and 27 days for sediment H2 and 2, 5, 15, 20, and 30 days for sediment S. After each residence time step, effluent was sampled and the core was completely flushed, replacing all pore water with new influent. Following the increasing residence time test, sediment S was subjected to temperature interval steps of 5°C and a 5 day residence time. The sediment initially kept at 5°C was brought to 15°C, the 11°C sediment was brought to 15°C and 20°C, the 25°C sediment was varied between 30°C and 50°C and the 60°C sediment was varied between 50°C and 80°C. A number of temperature steps was carried out in duplicate (30°C, 40°C, 50°C, 65°C) and in triplicate (5°C, 11°C, 15°C, 20°C). Following these tests, the sediment cores were kept again at the original temperatures of 5°C, 11°C, 25°C and 60°C for a period of two weeks before dismantling and sampling of sand for DNA extraction. Figure S2.1 shows the applied sampling schedule for both types of experiments.

S2.2 METHODS: MICROBIOLOGICAL ANALYSES

S2.2.1 DNA Extraction

DNA was extracted from sediment samples by adding 50 mL of autoclaved tap water to 2–5 g sediment samples. Subsequently the samples were sonicated for 2 min at 20 kHz in a Sonifier II W-250 and the liquid phase was collected. DNA was isolated from the liquid samples by filtration over a 25-mm polycarbonate filter (0.22 μm pore size, type GTTP; Millipore, the Netherlands). The filter was subsequently added to the phosphate and MT buffer of the FastDNA Spin kit for soil (Qbiogene) and stored at −20°C. DNA was isolated using the FastDNA Spin kit for soil according to the supplier's protocol. DNA was extracted following 40 days (effluent only) and 70 days (effluent and sand) of incubation of sediment H2, and at the end of the experiment (180 days) of the sand from sediment S.

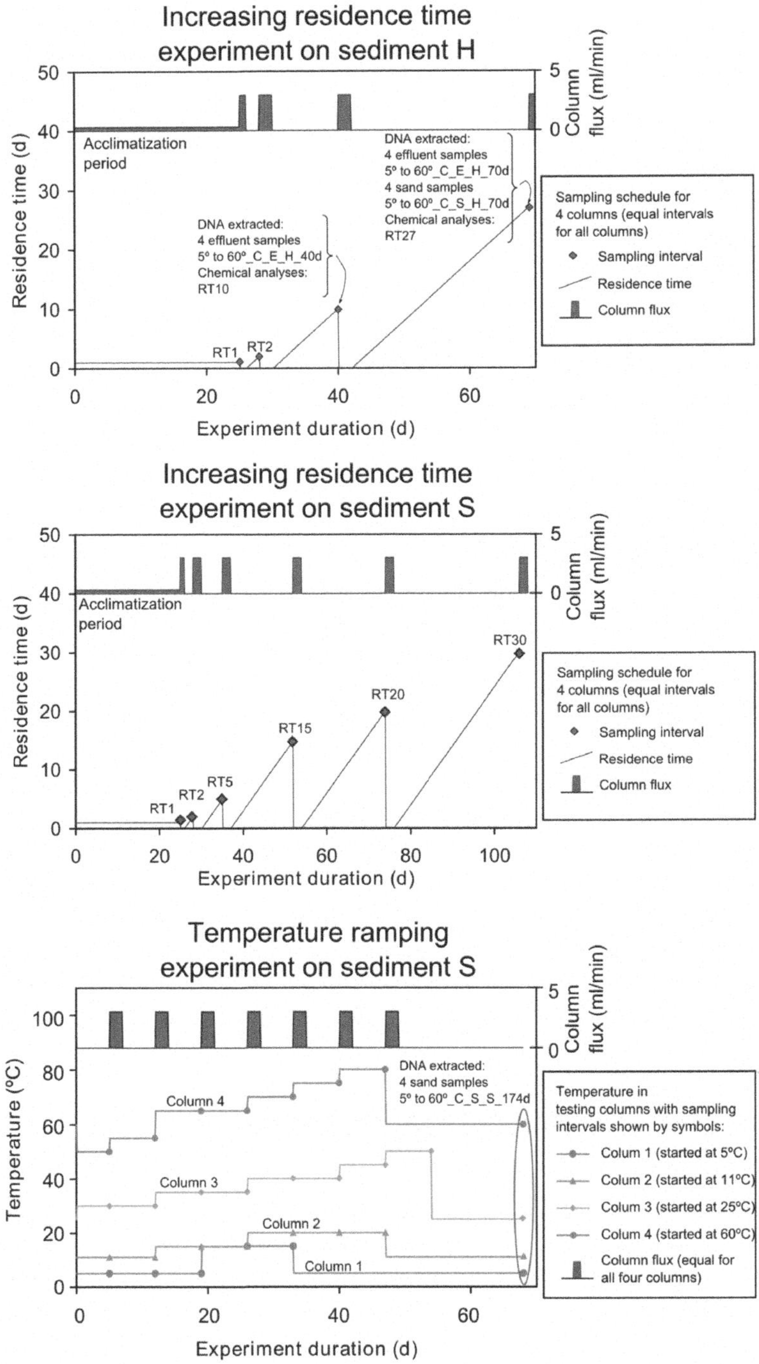

Figure S2.1 Sampling schedule for the two experiment types on sediment H2 and S. Sediment H2 was only used for the increasing residence time experiment, while sediments H and S were used for both types of experiments.

S2.2.2 T-RFLP fingerprinting of bacterial communities

The diversity and composition of the bacterial communities in the samples were determined by terminal restriction fragment length polymorphism (T-RFLP). For T-RFLP analysis, 16S rRNA genes were amplified using primers 8f-FAM (i.e., 6-carboxyfluorescein labeled) and 1392r, and the PCR amplification conditions. Fluorescently labeled PCR products (45 μL) were purified using a DNA Clean & Concentrator™-5 Kit according to the manufacturer's instructions (Zymo Research, USA). Digestion mixtures (20 μL) contained 5 U of restriction enzyme HhaI (Promega, The Netherlands), 2 μL of buffer C (Promega), 0.2 μL of BSA, and 5 μL of purified PCR product, and were incubated for 6 hours at 37°C. Subsequently, the digestion mixtures were purified as described above. Purified digestion products (5 μL) were mixed with 15 μL of loading buffer Hi-di formamide and 1 μL of internal standard GeneScan-1000 ROX (both Applied Biosystems, The Netherlands). After heating for 3 min at 95°C and subsequent cooling on ice, the fluorescently labeled terminal restriction fragments (T-RFs) were size separated on an ABI Prism 310 genetic analyzer (Applied Biosystems) in GeneScan analysis mode (Applied Biosystems). Cluster analysis based on Euclidean distance was performed with BioNumerics 6.0 software (Applied Maths, Belgium) by comparing the t-RFLP electropherograms using Pearson correlation as a distance measure (Figure S2.2).

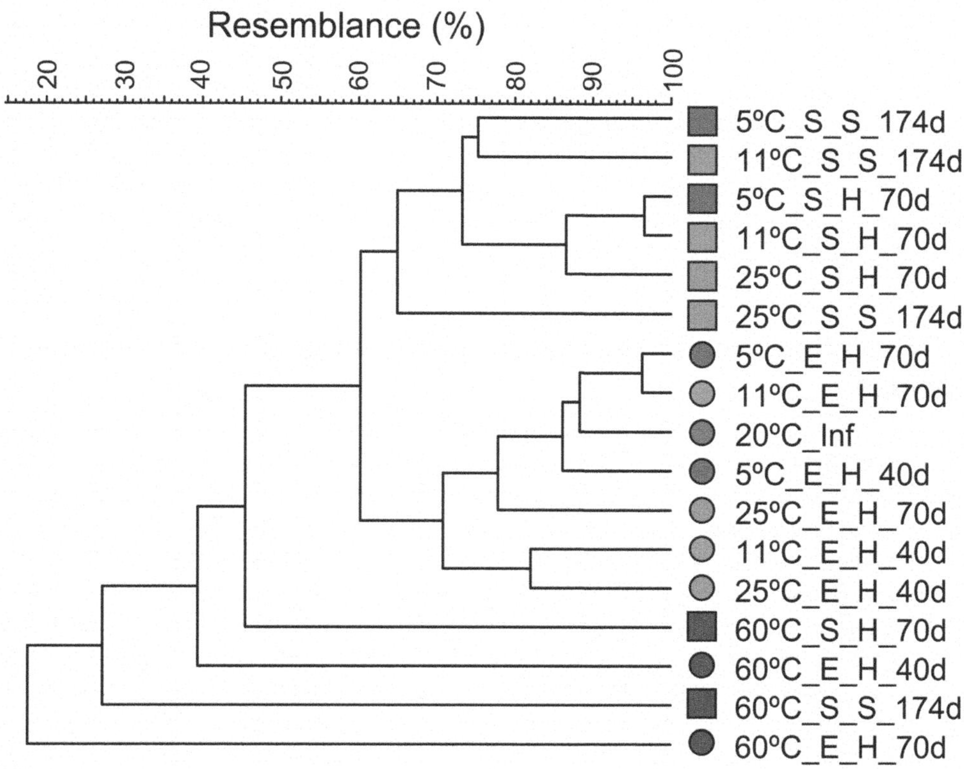

Figure S2.2 Dendrogram of the cluster analysis for T-RFLP data based on Euclidean distance using Pearson correlation as a distance measure.

S2.2.3 Amplicon preparation and pyrosequencing of bacterial 16S rRNA genes

Barcoded amplicon libraries of the small subunit ribosomal RNA gene hypervariable region V5-V7 were generated for each of the individual sample as described previously (Kraneveld *et al.* 2012) pooled and sequenced by means of the Genome Sequencer FLX Titanium system (Roche, Basel, Switzerland). The sequencing data was processed using QIIME (Quantitative Insights Into Microbial Ecology) version 1.5.0 (Caporaso *et al.* 2010). For further downstream analyses, barcodes and primer sequences were trimmed and low quality reads (reads containing ambiguous base calls, >1 error in the primer, >1 error in the barcode, >5 nt homopolymer sequence, the average quality score below 30, or a length <200 bp or >1000 bp) were removed from the analyses. Sliding window test (50 nt) of quality scores was enabled and sequences of low quality were truncated at the beginning of the poor quality window. The reads were denoised using Denoiser version 1.3.0 (Reeder & Knight, 2010). The de-noised reads were checked for chimeric sequences using UCHIME version 4.2.40 (Edgar *et al.* 2011). The results of the *de novo* and the reference-based approach were combined and reads marked as chimeric were removed.

The cleaned reads were clustered into Operational Taxonomic Units (OTUs) at a minimal sequence similarity of 97% using UCLUST Reference Optimal (Edgar, 2010). The representative sequence of each cluster was assigned a taxonomy using the Ribosomal Database Project (RDP) classifier (Cole *et al.* 2009). The minimum confidence was set at 0.8. For taxonomy assignment the SILVA rRNA database (Pruesse *et al.* 2007) was trimmed to span the targeted hypervariable regions V5–V7 as described by Brandt *et al.* (2012). The taxonomy assigned OTUs were aligned using PyNAST (Caporaso *et al.* 2009). For the alignment, the minimum aligned sequence length was set at 150 nt and the minimum percent identity at 75%. To allow comparisons among different samples, a randomly sub sampled dataset of 1670 reads per sample (the minimum number of reads per sample was 1673) was created. Raw sequences are available at NCBI short read archive under accession number SRP026337. Non-metric dimensional multiscaling after Bray Curtis similarities was performed in PAST version 2.14 (Hammer *et al.* 2001).

S2.2.4 Denaturing gradient gel electrophoresis (DGGE) of *archaeal* communities

16S rRNA gene fragments of *Archaea* were amplified by nested PCR, using first primers PRA46f and Univ0907r, followed by amplification with primers PARCH340f-GC and PARCH519r (second set). (Ovreås *et al.* 1997; Vetriani *et al.* 1999) A total volume of 25 μL was used in each PCR reaction, containing 0.4 μM forward and reverse primers; 0.4 mg mL^{-1} BSA (Bovine Serum Albumin – New England BioLabs, Leusden, The Netherlands); 12.5 μL Fidelitaq PCR Master Mix (2X) (USB Corporation, Cleveland, OH); 8.5 μL DNase and RNase-free water (MP Biomedicals, Solon, OH) and 1 μL template. For both PCRs an initial denaturation was performed at 94°C for 4 minutes; 35 cycles of 94°C for 30 s, 54°C for 1 min and of 72°C for 1 min; plus a final elongation step at 72°C for 5 min. PCR products were profiled by Denaturing Gradient Gel Electrophoresis (DGGE; Bio-Rad DCode Universal Mutation Detection System). Markers consisting of a mixture of 12 different bacterial 16S rRNA gene fragments were used alongside the samples to aid in normalization. Gels were 8% polyacrylamide (37, 5:1 acrylamide/Bis) with a denaturing gradient of 30/70%. Electrophoresis was performed in 1xTAE buffer during 4 hours at 60°C and 200 V. Gels were ethidium bromide stained, illuminated under a Vilber Lourmat (TCP-20-M) UV transilluminator and photographed.

Phylogenetic analysis was performed for bands excised from DGGE gels. DNA was eluted by incubation overnight at 4°C in 50 μL of TE buffer. One μL of the eluted DNA was used as a template for a PCR reaction with primers pARCH34f and pARCH519r without GC-clamp (Ovreås *et al.* 1997; Vetriani *et al.* 1999). Bidirectional sequencing of the product was carried out by Macrogen Europe (Amsterdam, The Netherlands). Contig assembly was done using DNA Baser software. The assembled sequences were compared with sequences deposited in GenBank using BLAST (Altschul *et al.* 1990).

S2.3 HYDROGEN THRESHOLDS FOR REDOX PROCESSES

The prevalence of different redox processes was evaluated using the energetics of the terminal electron accepting processes (TEAPs) (Hansen *et al.* 2001). In order to do this, we considered the four relevant TEAPs shown in Table S2.1 based on H_2 oxidation. H_2 is microbiologically produced in anoxic environments by fermentation of sedimentary organic matter, and subsequently used to reduce inorganic species.

Table S2.1 Reaction equations and thermodynamic data used to calculate the threshold H_2 concentrations for terminal electron accepting processes. Thermodynamic data taken from Stumm & Morgan (1996).

Reaction	ΔG^0 (kJ mol^{-1}) at 25°C	ΔH^0 (kJ mol^{-1})
$H_2 + 2FeOOH$ (goethite) $+ 4H^+ \leftrightarrow 2Fe^{2+} + 4H_2O$	−146.8	−198.7
$H_2 + 2FeOOH$ (lepidocrocite) $+ 4H^+ \leftrightarrow 2Fe^{2+} + 4H_2O$	−174.4	−198.5
$4H_2 + SO_4^{2-} + H^+ \leftrightarrow HS^- + 4H_2O$	−262.35	−235
$4H_2 + HCO_3^- + H^+ \leftrightarrow CH_4 + 3H_2O$	−229.41	−237.81

The general equation of writing the H_2 oxidizing reaction (with a stoichiometric coefficient for H_2 of (1) is given by (Hoehler *et al.* 1998; van Breukelen *et al.* 2003):

$$1H_2 + aOx + bH^+ \rightarrow cRed + dH_2O \tag{S2.1}$$

where aOx and cRed are the oxidized and reduced forms of a terminal electron acceptor. The free energy of this reaction, ΔG_r (kJ mol^{-1}) is given by:

$$\Delta G_r = \left(\Delta G^{0,T} + RT\mathrm{Ln}\left(\frac{[\mathrm{Red}]^c}{[\mathrm{Ox}]^a[\mathrm{H}^+]^b}\right)\right) - RT\mathrm{Ln}[\mathrm{H}_2] \tag{S2.2}$$

where $\Delta G^{0,T}$ are the Gibbs free energy of the reaction at given temperature $T(K)$ under standard conditions, which can be corrected using the Gibbs–Helmholtz equation, and R is the gas constant. Eq S2.2 can be used to derive the threshold or minimum H_2 activity for the redox reactions shown in Table S2.1 to proceed, assuming a minimum free energy ΔG_r required for biochemical reactions to proceed. We used a minimum free energy of −7 kJ mol^{-1} H_2 based on Jakobsen *et al.* (1998) and van Breukelen *et al.* (2003). The latter can be considered to equal the free energy for ATP formation and implies that the redox process must yield at least enough energy to produce ATP (LaRowe & Van Cappellen, 2011; LaRowe *et al.* 2012).

S2.4 MICROBIOLOGICAL RESULTS

Table S2.2 Available as separate Excel Spreadsheet (supplementary information available on-line), Characteristic Operating Taxonomical Units (OTUs) present in sediment (gray columns) and water (white columns) obtained from column experiments conducted at various temperatures. Three features are indicated, separated by lines. The top part in the table shows OTUs present at in total 10 or more counts in samples derived from columns conducted at 60°C, while these OTUs were not observed at lower temperature. The middle part shows OTUs indicative of sulfate reduction. The lower part shows OTUs indicative of iron reduction. Per feature, OTUs are sorted on basis of relative abundance, with the highest abundances (over all samples) on top. Classification is shown up to genus level. Physiology was derived from general, scientific literature (e.g., Bergey's manual of Systematic Bacteriology, The prokaryotes, International Journal of Systematic and Evolutionary Microbiology), to indicate characteristics observed for members of a particular genus: Sp, sporeforming; Th, moderate thermophilic (40–65°C); Sulf, sulfate-reducing; Fe, iron-reducing. The percentage occurrence of each OTU in a particular sample is given. It should be noted that OTUs only found in samples from the 60°C experiment and indicative of sulfate- or iron-reduction, are not repeated in the middle and lowest part, respectively, but are included in the total abundances of sulfate- and iron-reducers. Raw sequences are available at NCBI short read archive under accession number SRP026337.

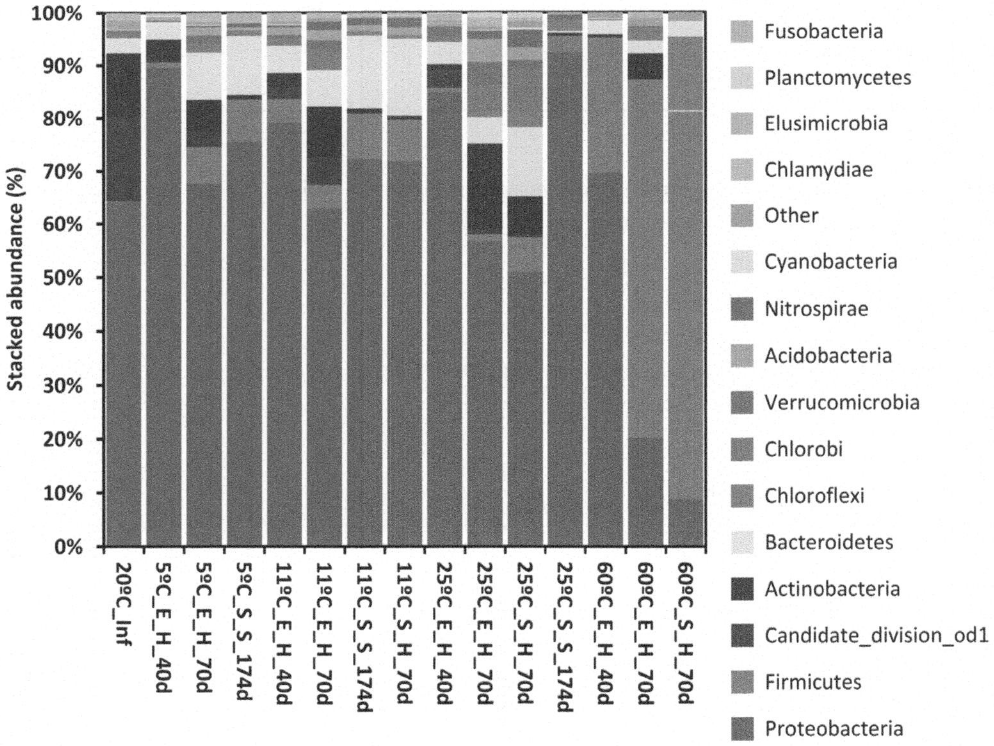

Figure S2.3 Relative abundances of reads (in percentage) attributed to phyla in sediment and water obtained from column experiments conducted at various temperatures.

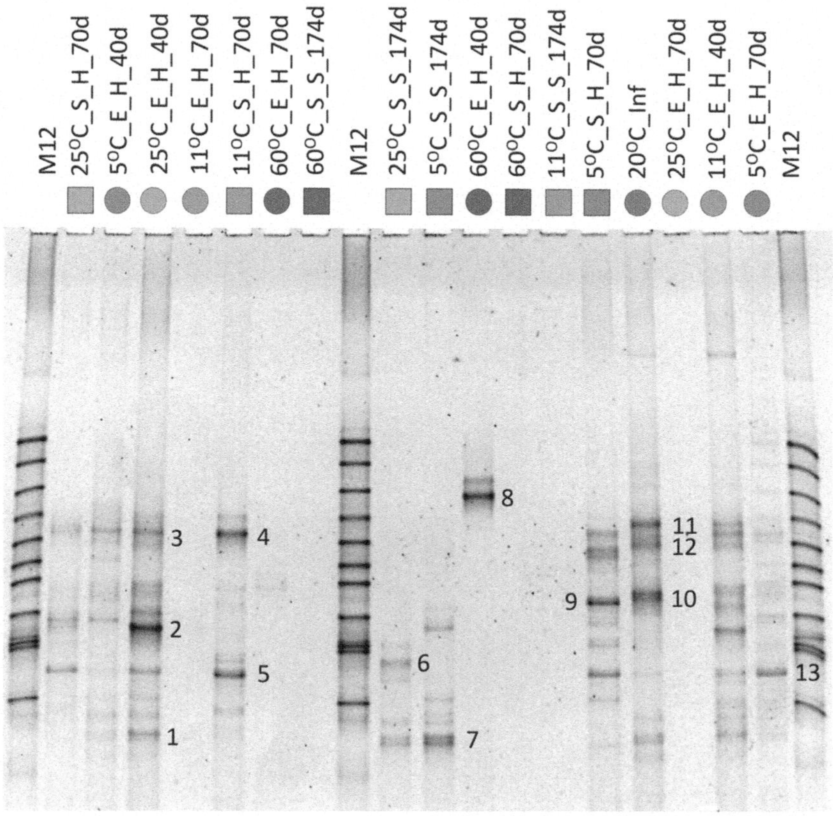

Band	Closest hit	Accession number	Similarity
1	Haloferax sp. Bej51	GU361125.1	82%
2	Haloferax sp. Bej51	GU361125.1	83%
3	Methanococcus vannielii SB	NR_074175.1	81%
4	Marine crenarchaeote RS.Sph.033	DQ097307.1	91%
5	Methanothermus fervidus DSM 2088	NR_102926.1	84%
6	Methanothermus fervidus DSM 2088	NR_102926.1	84%
7	Methanocaldococcus indicus strain SL43	NR_028861.1	84%
8	Nitrosopumilaceae archaeon MY1	HQ331116.1	97%
9	Vulcanisaeta moutnovskia 768-28	NR_102947.1	82%
10	Candidatus Methanomethylophilus alvus Mx1201	CP004049.1	89%
11	Haloarcula sp. PCR16SARA2	KC312960.1	88%
12	Methanosaeta harundinacea	NR_102896.1	78%
13	Haloferax sp. Bej51	GU361125.1	83%

Figure S2.4 DGGE profiles of archaeal 16S rRNA gene fragments (30–70% denaturing gradient) obtained from sediment and water sampled from column experiments conducted at various temperatures. An empty lane indicates that no PCR product was obtained. Numbers refer to bands that have been cut out and sequenced. Closest hits of cultured species, with accession numbers and percentage similarity are indicated in the table.

S2.5 KINETIC AND THERMODYNAMIC PARAMETERS OF SULFATE REDUCTION REPORTED IN LITERATURE

Table S2.3 Literature values for sulfate reduction kinetics and thermodynamics in marine and aquifer sediments. If two samples are available, the range is shown. With more samples the average and standard deviation are shown.

Sediment type or experiment with note to reference	$T_{in\text{-}situ}$ (°C)	T_{opt} (°C)	T_{max} (°C)	$r_{in\text{-}situ}$[a] (M s^{-1})	r_{opt}[a] (M s^{-1})	E_a (kJ mol^{-1})	Q_{10}
Our experiments							
A. Increasing residence time							
Sediment H2 (5–25°C)	11	n.a.	n.a.	4.8×10^{-12}	n.a.	115 ± 267	7.0
Sediment S (5–25°C)	11	n.a.	n.a.	4.3×10^{-11}	n.a.	54 ± 26	2.4
B. Temperature ramping							
Sediment S (5–40°C)	11	40	47	4.3×10^{-11}	$4.8 \pm 6.1 \times 10^{-10}$	41 ± 19	1.9
Sediment S (50–70°C)	11	70	>80	n.a.	$5.3 \pm 11 \times 10^{-10}$	80 ± 44	3.2
Literature values for aquifer sediments							
Rømø aquifer, Holocene eolian sand[b]				$1.6\text{–}140 \times 10^{-12}$			
Tuse Næs, Pleistocene glacial sand[b]				$5\text{–}12 \times 10^{-12}$			
Data compilation of others[b]:							
Cretaceous sands and carbonates (Black Creek,US)				$2.2\text{–}48 \times 10^{-16}$			
Tertiary limestone (Florida, US)				3.2×10^{-15}			
Cretaceous sandstone (Fox Hills, US)				6.3×10^{-15}			
Pleistocene sand (Führberg, DE)				4.4×10^{-13}			
Pleistocene sand (Sturgeon Falls, CA)				4.1×10^{-13}			
Pleistocene sand (Bocholt, DE)				7.3×10^{-13}			
Reactive permeable aquifer barrier (pea gravel & compost)[c]				1.3×10^{-9}		40	1.8
Literature values for marine and estuarine sediments (all Holocene)							
Shelf & slope sediments[d] (muds with organic carbon, 1–7%OC), $n = 10$	3.2	22.6	33	$2.3 \pm 3.1 \times 10^{-10}$	$1.5 \pm 1.7 \times 10^{-9}$	34 ± 11	2.5 ± 0.4
Estuarine silty clay and sand[e] (0.8–4%OC), $n = 6$	3–27			$4.5 \pm 4.1 \times 10^{-9}$	$7.2 \pm 7.4 \times 10^{-9}$	51 ± 19	2.3 ± 0.6
Marine clay/mud[f] Arctic	2	22–32	40	$0.9\text{–}5.1 \times 10^{-10}$	1.3×10^{-8}	50	2.1
Moderate climate	3–20	30–36	>40	$0.2\text{–}7.6 \times 10^{-10}$	3.1×10^{-8}	81	3.2

Literature values for geothermal sediments							
Hydrothermal vent sediment at[g]:							
2.2×10^7 Pa (in situ) (73–100°C)	80	100	105	1.5×10^{-8}	1.3×10^{-7}	115	5.3
4.5×10^7 Pa (73–95°C)	80	95	105		1.6×10^{-7}	96	4.0
Yellowstone geothermal Park[h]:							
Obsidian pool, 35, 65, 85°C	85			2.9×10^{-10}		17	1.2
Black sediment pool, 30, 40, 70°C	70			4.9×10^{-9}		38	1.5
Mushroom spring, 30, 60, 70°C	60			6.4×10^{-9}		119	3.8
Guaymas Basin hydrothermal vent[i] (clays/silts):							
sample A (53–63°C)	25	63	67		3.2×10^{-9}	254	40
sample B (70–77°C)	70	77	83		1.9×10^{-9}	334	127
sample C (60–83°C)	75	83	90		6.9×10^{-9}	102	4.4
Literature values for incubated experiments at both low and high temperatures							
Marine Arctic sediment[j]:							
Psychrotolerant range (0–22°C)	–2 – +4	22	32	$1 - 15 \times 10^{-10}$	1.4×10^{-9}	71	2.8
Thermophilic range (47–56°C)	–2 – +4	56	64	$1 - 15 \times 10^{-10}$	8×10^{-8}	98	4.1
Marine moderate sediment[k]:							
Mesophilic range (0–40°C)	0–15	31	40	5.8×10^{-10}	5.8×10^{-9}	75	3.0
Thermophilic range (45–58°C)	0–15	58	65		5.1×10^{-9}	199	17.9
Geothermal lacustrine sands[l]:							
Mesophilic range (20–50°C)	60–101	45			1.2×10^{-9}		
Thermophilic range(50–82°C)	60–101	65	82		1.2×10^{-9}	114	3.3

Notes: [a]Marine SRR are often reported in nmol cm^{-3} day^{-1} determined by 35S radiotracer method where SR is determined per cm^3 sediment. These values are converted to a SR in M s^{-1} using the published porosity for the sediment used. If no porosity is given, a porosity of 0.5 was assumed.

[b]Jakobsen and Postma (1994).

[c]Benner *et al.* (2002).

[d]Sawicka *et al.* (2012).

[e]Marvin-DiPasquale and Capone (1998).

[f]Robador *et al.* (2009).

[g]Kallmeyer & Boetius, 2004), Ea calculated with SRR and temperatures shown in Table 1 in Kallmeyer and Boetius (2004).

[h]Roychoudhury (2004).

[i]Jørgensen *et al.* (1990), Ea calculated from SRR and temperatures shown in Figure 6 in Jørgensen *et al.* (1990).

[j]Hubert *et al.* (2009, Hubert *et al.* (2010) Calculated from SRR and temperatures shown in Figure 1 in Hubert *et al.* (2009).

[k]Isaksen *et al.* (1994), Calculated from SRR and temperatures shown in Figure 1B in Isaksen *et al.* (1994).

[l]Elsgaard *et al.* (1994).

Appendix 3

Supplementary information

This appendix contains detailed descriptions of the (1) experimental setup and sediment characterization, (2) K-feldspar weathering and cation exchange modeling, (3) summary of optimized parameters, (4) the literature survey of SCM enthalpies, and (5) additional DAFT model results. The PHREEQC models and scripts can be downloaded free of charge from the supplementary information for the article this chapter was based on: http://pubs.acs.org.

S3.1 METHODS

S3.1.1 Sediment sampling and characterization

The sediment samples were collected using cores from boreholes drilled with the percussion drilling method. The sediment sample was stored in a second PVC liner filled with water sparged with N_2 gas and spiked with an anti-oxidant ($NaHSO_3$) in order to avoid permanent gas use in the glovebox during storage.

A combination of geochemical analyses was used to characterize the sediments (Table S3.1). The measured geochemical parameters were also used to calculate relevant reactive capacity properties: pyrite content and non-pyrite reactive iron (iron oxides) content based on empirical relations derived for Dutch sediments (including the geological formations used here) reported by Griffioen *et al.* (2012).

S3.1.2 Experimental setup

The cores with sediment were placed in the experimental set-up shown in Figure S3.1 and kept at temperatures of 5, 11, 25, and 60°C. The entire setup was placed in a 11°C climate room, the 5°C core was additionally cooled with a circulation cooler, and the 25 and 60°C cores were heated with heating mats (Mechaheat custom made). Influent water used to replace the sampled water was kept at a constant excess pressure of 1 bar (compared to atmospheric pressure) with a mixture of N_2/CO_2 gas. The CO_2 fraction was adjusted to the value under field conditions. A pressure relieve valve maintained an excess pressure within the sediment cores of 1 bar. The sediment cores were flushed for 25 days at 0.33 ml/min resulting in a water residence time of 1 day. A full description of the experimental methods is provided in Bonte *et al.* (2013).

S3.1.3 Chemical analyses

Major cations were analyzed with inductively coupled plasma optical emission spectrometry (ICP-OES, Varian 720 ES-axial), anions with conductometry after ion chromatography (Dionex DX-120 IC equipped with IonPac AS14 column), dissolved organic carbon (DOC) was measured using high temperature catalytic oxidation (Shimadzu Model TOC-500A analyzer), and methane with gas chromatography.

Table S3.1 Geochemical and hydraulic characteristics of the sediment used in the column experiments. A full description on geochemical analyses is provided in Bonte *et al.* (2013).

Location			**Scherpenzeel**
Sampling depth (m-surface level)			**34–36**
Measured parameters	**Method**	**Unit**	
Reactive organic matter (LOI330)	Thermogravimetric analysis: Loss on ignition (LOI) at 330, 550 and 1000°C	% d.w.	0.06
Bulk organic matter (LOI550)[b]			0.14
Carbonate content (LOI1000)			0.74
Clay content	Laser grain size analyses		1.34
Median grain size (d_{50})		μm	310
Uniformity coefficient (d_{60}/d_{10})		–	2.1
S	Carbon Sulphur (CS) Combustion Analyzer (LECO)	% d.w.	0.05
Minerals present	X-Ray diffraction (XRD)		quartz, K-feldspar, albite, ferrian clinozoisite, calcite, dolomite
Cr_2O_3	X-ray fluorescence (XRF)	%	0.001
Fe_2O_3			0.59
Al_2O_3			3.36
As	$LiBO_2$ fusion followed by	ppm	3.7
Mo	HNO_3 digestion and inductively coupled plasma – mass spectrometry (ICP-MS)		0.1
V			<8
Be			<1
Calculated parameters	**Equation**	**Unit**	
Bulk density, ρ_b	Weight/volume	g/cm³	2.12
Porosity, n	$n = (\rho_{bulk} - \rho_{H_2O})/(\rho_{solid} - \rho_{H_2O})$		0.32
Pyrite[b]	$0.5 * S * M_{FeS_2}/M_S$	% d.w.	0.09
Reactive iron (iron oxides)[b]	$2 * M_{Fe}/M_{Fe_2O_3} * (Fe_2O_3 - (0.225 * Al_2O_3 - 0.91\%)) - 0.5 * S * M_{Fe}/M_S$		0.48

[a]corrected for structural water loss from clay according BOM = LOI550 – 0.7 × Clay% (Breeuwsma, 1987).
[b]based on Griffioen *et al.* (2012) with M_{FeS_2}, M_{Fe}, M_S, $M_{Fe_2O_3}$ being the molar masses of FeS_2, Fe, S, and Fe_2O_3, respectively (55.8, 32.1, 159.7), and Fe_2O_3 and Al_2O_3 the % d.w. of iron oxides and aluminium oxides in sediment determined by XRF, and S being the content of sulphur determined with the CS combustion analyser.

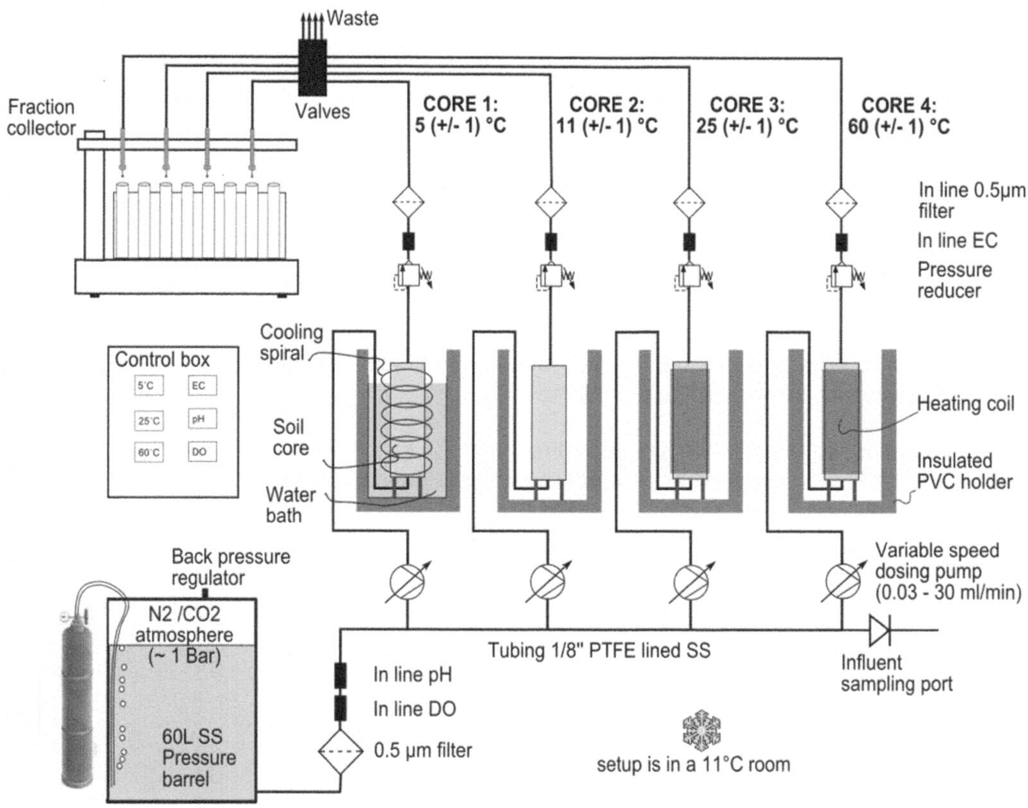

Figure S3.1 Experimental setup used for anoxic column experiments. A full description on experimental methods is provided in Bonte *et al.* (2013).

S3.2 ANALYSIS OF CONSERVATIVE BREAKTHROUGH TESTS

The results of the breakthrough tests were analysed using the analytical model CXTFIT (Toride *et al.* 1995). This model is based on the advection-dispersion equation and estimates the main transport parameters by fitting the experimental data using a nonlinear least-squares parameter optimization method (Toride *et al.* 1995). Assuming the tracer is fully conservative, we used the model to calculate dispersivity and effective porosity. Figure S3.2 and Table S3.2 present the results of the breakthrough tests (BTT) of the four columns and the derived porosity (*n*) and dispersivity (*d* in cm), respectively. Also shown in Table S3.2 are R^2 and the SE of the model fit.

Klotz *et al.* (1980) found dispersivity values ranging between 0.2 to 23.5 cm depending on both the median grain size and heterogeneity as expressed by the uniformity coefficient (U, the ratio of d60 and d10, respectively the 60% and 10% sieve passing, values of our sand shown in Table S3.1). Another study by Perfect *et al.* (2002) found dispersivity to range between 0.1 and 19 cm with most values between 4 and 7 cm. Results in this study depended as well on sediment texture where finer sediments generally yielded higher dispersivities. Comparing the dispersivities with our values shows that while our dispersivity values are at the high end of what can be expected in column tests, they are not the result of preferential flow paths.

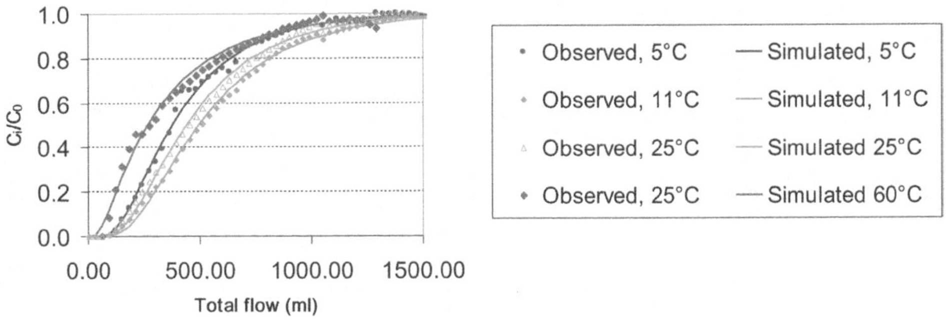

Figure S3.2 Results from breakthrough tests at four temperatures. Solid lines are CXTFIT model results to determine dispersivity, symbols are observations.

Table S3.2 Summarised results from simulated breakthrough tests: dispersivities and porosities determined with CXTFIT, compared with the porosity calculated from bulk weights.

Column	R^2	Dispersion coefficient, D (cm^2/min)	St. error	Dispersivity, α (cm)	St. error	Total porosity (n_{total})
Sediment C						
5°C	0.9956	3.76	0.041	10.4	0.11	0.28
11°C	0.9995	2.53	0.010	9.2	0.04	0.31
25°C	0.9980	3.03	0.019	9.8	0.06	0.33
60°C	0.9896	13.01	0.099	23.8	0.18	0.34
average	1.00	5.58	0.000	13.3	0.10	0.32

S3.3 ADDITIONAL INFORMATION ON SIMULATED CATION EXCHANGE

Table S3.3 Reactions and control parameters (log_K and ΔH_r values) to simulate cation-exchange, initial values (PHREEQC database), optimized values (average and standard error), and reported literature values or ranges.

	Log *K*			ΔH_r (kJ mol^{-1})	
	Initial[a]	Optimized	Literature range compiled by Karlsen, *et al.* (2012)	Initial[a]	Optimized
$Na^+ + X^- \leftrightarrow NaX$	0			0	
$K^+ + X^- \leftrightarrow KX$	0.7	1.4 ± 0.1	0.2–2.6	−4.3	−19 ± 1
$Ca^{2+} + 2X^- \leftrightarrow CaX_2$	0.8		0.3–2.4	7.2	
$Mg^{2+} + 2X^- \leftrightarrow MgX_2$	0.6		0.2–1.6	7.4	
$Mn^{+2} + 2X^- = MnX_2$	0.52			0	
$Fe^{+2} + 2X^- = FeX_2$	0.44			0	
$Sr^{2+} + 2X^- \leftrightarrow SrX_2$	0.91	0.8 ± 0.2		5.5	0.55 ± 1.2

[a]based on PHREEQC database, original references are in database.

S3.4 ADDITIONAL INFORMATION ON THE KINETIC DISSOLUTION OF K-FELDSPAR

The rate equation (6.5) in the main paper for kinetic dissolution of K-feldspar applies to a specific rate in keq · m^2s^{-1} which is multiplied by the ratio of mineral area and volume of water in contact with this area. The water volume, V, (l) in contact with feldspar within the sediment (kg) is calculated with:

$$V = \frac{M_t n}{\rho_{\text{solid,total}}(1-n)} \tag{S3.1}$$

where M_t is the total mass of dry sediment (usually taken as 1 kg), and $\rho_{\text{solid,total}}$ is the average dry density of all solids. A porosity of 0.32 and solid grain density of 2.65 kg/dm³ yield $V = 0.18$ dm³/kg. The initial reactive surface area of the mineral under consideration is usually determined in laboratory experiments with gas adsorption, a so called 'BET' measurement (Brunauer *et al.* 1938). For natural or mixed sediments like the sediments used here, this method was not feasible. An indication of the initial reactive surface area was achieved from the geometric surface area, A_{geo}, similar as done by Sverdrup (1990):

$$A_{\text{BET}} = A_{geo}\lambda = \frac{6M_i}{\rho_{\text{solid,i}}\,\phi\,\bar{d}}\lambda \tag{S3.2}$$

where M_i is the mass of mineral i involved (kg) which has to be related to the volume of water in contact with this mineral, ρ_{solid} is the dry density of the mineral (ranging between 2.55 to 2.76 kg/dm³; taken as $\rho = 2.65$ kg/dm³, the specific density of quartz), ϕ is the sphericity of the grains (ranging between 0.85 to 0.9 (Sverdrup, 1990), taken as 0.87), d is the average grain size (310 µm based on the median grain size Table S3.1), and λ (surface roughness) is defined as the ratio between the BET surface area and the geometric surface area (Helgeson, 1971). λ unfortunately is very poorly constrained but generally increases with age, ranging from $\lambda = 1$ for synthetic freshly prepared minerals to around $\lambda = 400$ for field minerals deposited >10 (Perfect *et al.* 2002) years ago (White *et al.* 2003). This parameter was allowed to vary allowing for model calibration. An initial value of $\lambda = 40$ was applied.

Table S3.4 Parameters used in kinetic modeling of K-feldspar dissolution. pk (–log(k)) and n values are defined for 25°C. E_a = activation energies (values from on Sverdrup, 1990).

	pk (25°C) (keq m^{-2} s^{-1})	n	E_a (kJ mol^{-1})
H^+	12.5	0.5	60.0
H_2O	15.3	1	35.0
OH^-	14.2	0.3	32.1
CO_2	14.6	14.6	59.3
DOC	13.9	0.4	24.0

S3.5 ADDITIONAL INFORMATION ON THE METHODOLOGY FOR SENSITIVITY AND UNCERTAINTY ANALYSIS

In order to find the optimized solution of the model, PEST calculates the Jacobian matrix during each optimization iteration. Part of the solution methodology employed by PEST involves calculation the 'composite sensitivity' of each parameter. This is used by PEST to derive the parameter set for the next iteration. The composite sensitivity gives an indication of the sensitivity of the model fit to each of the parameters. In order to compare sensitivies among different model parameters, the composite sensitivities are multiplied by the absolute parameter values to give the relative sensitivities. The relative sensitivities are shown in Table S3.5.

Using PEST to calculate the predictive uncertainty involves running a 'composite model' of calibration and predictive models. The composite model will use a parameter set and run the calibration model to determine the calibration error, while the predictive model is run to find the prediction value. PEST will use the composite model to find a parameter set which keeps the calibration error within a reasonable range from the optimized solution, while maximizing or minimizing the prediction using the numerical method developed by Vecchia and Cooley (1987). Highly parameterized models often are non-unique in the sense that multiple parameter sets can deliver the same minimum objective function, and in uncertainty mode, PEST will aim to find the maximum value for a given parameter, while keeping the increment in the calibration objective function sufficiently small. The calibration objective function for the optimized model, Φ, may vary within a reasonable range, δ, to find the uncertainty limits. The allowable increment in the objective function, δ, is determined by the probability level $1-\alpha$ required, and given by (Doherty, 2010):

$$\delta = n\sigma^2 F_\alpha(n, m - n) \tag{S3.3}$$

where n is the number of parameters to be optimized, m is the number of observations in the dataset, F denotes the F distribution with n and $m - n$ degrees of freedom, and s is given by (Doherty, 2010):

$$\sigma = \Phi/(m - n) \tag{S3.4}$$

where Φ is the minimized value of objective function.

S3.6 ADDITIONAL RESULTS

S3.6.1 Surface complexation

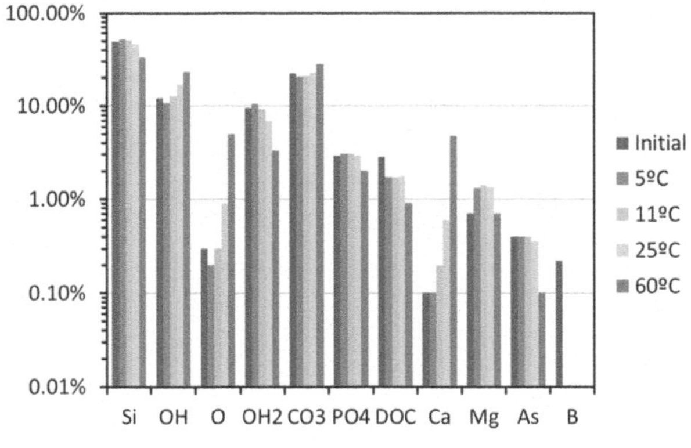

Figure S3.3 Surface speciation at the last cell calculated at the start (equal for all four testing temperatures) and at the end of the column experiment for the four testing temperatures.

S3.6.2 K-feldspar dissolution and cation exchange

Overall, there is a good agreement between simulated and observed concentrations of K and Si (Figure 6.1) suggesting that the chosen combination of K-feldspar dissolution, surface complexation (Si), and cation-exchange (K) can explain the enrichment of these elements with higher temperature. The relative contribution of these individual processes was investigated with the model by adding these processes one by one in consecutive runs (in Figure S3.4). This shows that the high potassium concentrations of the first pore flushes at 60°C is primarily explained by K release from the cation-exchanger. With progressive flushing, desorption from the cation exchanger decreases and the influent becomes sub-saturated with respect to K-feldspar resulting in its dissolution. Silica shows a similar pattern, but in this case reducing surface complexation of silica (thus desorption) is initially the main contributor during the first pore flushes followed by K-feldspar dissolution.

The optimized reaction enthalpy for K/Ca cation exchange ($\Delta H = -45$ kJ mol^{-1}, Table S3.4 & S3.5) was nearly a factor three larger than the value (–15.9 kJ mol^{-1}) from the PHREEQC database. A possible explanation for this discrepancy is that potassium may be released from K-feldspar through exchange of potassium by H^+ as described by Wollast (1967) a process also suggested by Holm *et al.* (1987) to occur at a high temperature heat storage system (90–115°C). This means that the K-feldspar dissolution is incongruent, and the mechanism is essentially a combination of dissolution and cation exchange, which is in our model mimicked by simulating both ion exchange and dissolution.

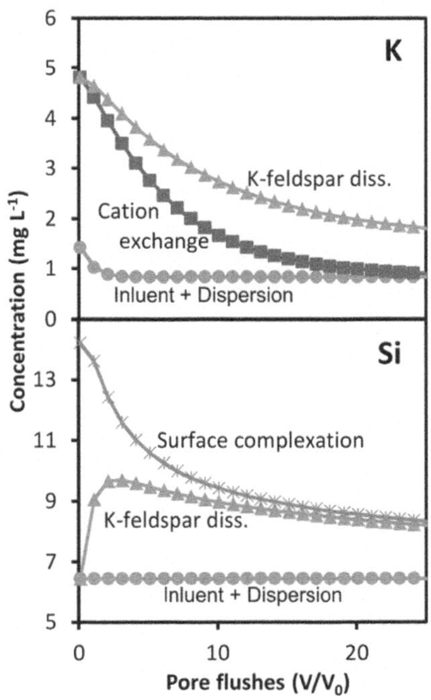

Figure S3.4 Contribution of cation exchange, K-feldspar dissolution, and surface complexation to simulated potassium (top panel) and silica (bottom panel) concentrations at 60°C. Concentrations were calculated by progressively adding processes to the PHREEQC model, in the order: (i) influent and dispersion, (ii) cation-exchange, (iii) K-feldspar dissolution, and (iv) surface complexation.

S3.7 OTHER INFORMATION

Table S3.5 Overview of all optimized parameters with standard error and relative sensitivity.

Reaction	Parameter	Optimized value ± S.e.	Relative sensitivity (%)
$K^+ + X^- \leftrightarrow KX$	$\log K_{XK \leftrightarrow K+X}$	1.4 ± 0.1	2.3%
	$\Delta H_{XK \leftrightarrow K+X}$	-19 ± 1	1.6%
$Sr^{2+} + 2X^- \leftrightarrow SrX_2$	$\log K_{XSr \leftrightarrow K+Sr}$	0.8 ± 0.2	0.4%
	$\Delta H_{XSr \leftrightarrow K+Sr}$	0.55 ± 1.2	0.04%
$\equiv FeOH + H_3AsO_3 \leftrightarrow \equiv FeH_2AsO_3 + H_2O$	$\log K_{\equiv FeH2AsO3}$	4.9 ± 0.2	28.3%
	$\Delta H_{\equiv FeH2AsO3}$	-43 ± 4	11.3%
$\equiv FeOH + H_3BO_3 \leftrightarrow \equiv FeH_2BO_3 + H_2O$	$\log K_{\equiv FeH2BO3}$	3.4 ± 0.4	1.1%
	$\Delta H_{\equiv FeH2BO3}$	-25 ± 20	0.2%
$\equiv FeOH + DOCA^- + H^+ \leftrightarrow \equiv FeDOCA + H_2O$	$\log K_{\equiv FeRDOC}$	11.6 ± 0.3	8.2%
	$\Delta H_{\equiv FeRDOC}$	-14 ± 30	0.2%
$\equiv FeOH + DOCB^- + H^+ \leftrightarrow \equiv FeDOCB + H_2O$	$\log K_{\equiv FeBDOC}$	12.6 ± 0.3	15.2%
	$\Delta H_{\equiv FeBDOC}$	-97 ± 20	2.7%
$\equiv FeOH + H_4SiO_4 \leftrightarrow \equiv FeH_3SiO_4 + H_2O$	$\log K_{\equiv FeH3SiO4}$	4.1 ± 0.3	16.2%
	$\Delta H_{\equiv FeH3SiO4}$	-21 ± 4	4.0%
$\equiv FeOH + PO_4^{3-} + 2H^+ \leftrightarrow \equiv FeHPO4^- + H_2O$	$\log K_{\equiv FeHPO4-}$	25 ± 0.6	15.2%
	$\Delta H_{\equiv FeHPO4-}$	-59 ± 6	4.6%
$\equiv FeOH + CO_3^{2-} + H^+ \leftrightarrow \equiv FeCO_3^- + H_2O$	$\Delta H_{\equiv FeCO3-}$	-49 ± 10	1.6%
$\equiv FeOH + CO_3^{2-} + 2H^+ \leftrightarrow \equiv FeHCO_3 + H_2O$	$\Delta H_{\equiv FeHCO3}$	-32 ± 7	2.2%
$\equiv FeOH + Mg^{2+} \leftrightarrow \equiv FeOMg^+ + H^+$	$\Delta H_{\equiv FeOMg+}$	1.4 ± 20	0.01%
$\equiv FeOH + Ca^{2+} \leftrightarrow \equiv FeOCa^+ + H^+$	$\Delta H_{\equiv FeOCa+}$	57 ± 30	0.8%
$r_{0,Kfsp} = \frac{\lambda A}{V} k_{sum}$	λ	96 ± 6	1.7%

Table S3.6 Enthalpies derived from literature for cation sorption to iron, aluminium, manganese oxides and related oxides (rutile, NiO).

Solute	Medium	ΔH_r (kJ mol^{-1})	Type	Source
Earth alkaline metals				
Ca	HFO	34.8	Correlation model	Trivedi and Axe (2001)
Ba	HFO	41.8	Correlation model	Trivedi and Axe (2001)
Sr	HFO	43.8	Correlation model	Trivedi and Axe (2001)
Mg	HFO	80.2	Correlation model	Trivedi and Axe (2001)
Ra	HFO	85.7	Correlation model	Trivedi and Axe (2001)
Be	HFO	111.6	Correlation model	Trivedi and Axe (2001)
(Post) transition metals				
Cd	HFO	96.0	Correlation model	Trivedi and Axe (2001)
Cd	Goethite	13	Experiment	Johnson (1990)

Table S3.6 Enthalpies derived from literature for cation sorption to iron, aluminium, manganese oxides and related oxides (rutile, NiO) (*Continued*).

Solute	Medium	ΔH_r (kJ mol^{-1})	Type	Source
Cd	Goethite	endothermic	Experiment	Mustafa *et al.* (2006)
Cd	Goethite	endothermic	Experiment	Gräfe *et al.* (2007)
Cd	Goethite	27–34	Experiment	Angove *et al.* (1999)
Cd	Polyacrylamide-grafted iron(III) oxide	16.45	Experiment	Manju *et al.* (2002)
Cd	Iron oxides	24.31	Experiment	Mustafa *et al.* (2010)
Cd	Hematite	–7	Experiment	Fokkink *et al.* (1990)
Cd	Hematite nanoparticles	27.21	Experiment	Shipley *et al.* (2013)
Cd	Rutile	5	Experiment	Fokkink *et al.* (1990)
Cd	Rutile	20	Model	Fokkink *et al.* (1992)
Cd	Rutile	15	Experiment	Fokkink *et al.* (1992)
Cd	NiO	29.4	Experiment	Mahmood *et al.* (2011)
Pb	HFO	173.4	Correlation model	Trivedi and Axe (2001)
Pb	Goethite	endothermic	Experiment	Rodda *et al.* (1993)
Pb	Polyacrylamide-grafted iron(III) oxide	42.66	Experiment	Manju *et al.* (2002)
Pb	Hematite nanoparticles	32.6	Experiment	Shipley *et al.* (2013)
Pb	Al2O3-supported iron oxide	25.7	Experiment	Huang *et al.* (2007)
Pb	MnO_2	9.35	Experiment	Zou *et al.* (2006)
Co	HFO	80.2	Correlation model	Trivedi and Axe (2001)
Co	Goethite	25–27	Experiment	Angove *et al.* (1999)
Co	NiO	32.7	Experiment	Mahmood *et al.* (2011)
Cu	HFO	78.3	Correlation model	Trivedi and Axe (2001)
Cu	Natural iron oxide-coated sand	52.32	Experiment	Boujelben *et al.* (2009)
Cu	Goethite	endothermic	Experiment	Rodda *et al.* (1993)
Cu	Hematite nanoparticles	48.4	Experiment	Shipley *et al.* (2013)
Cu	MnO_2	7.2	Experiment	Zou *et al.* (2006)
Cu	Waste iron oxide	9.2	Experiment	Huang *et al.* (2007)
Zn	HFO	112	Experiment	Trivedi *et al.* (2014)
Zn	HFO	127	Experiment	Trivedi *et al.* (2014)

(*Continued*)

Table S3.6 Enthalpies derived from literature for cation sorption to iron, aluminium, manganese oxides and related oxides (rutile, NiO) (*Continued*).

Solute	Medium	ΔH_r (kJ mol^{-1})	Type	Source
Zn	HFO	128	Experiment	Trivedi *et al.* (2014)
Zn	HFO	130	Experiment	Trivedi *et al.* (2014)
Zn	HFO	87.2	Correlation model	Trivedi and Axe (2001)
Zn	HFO	95.1	Experiment	Trivedi and Axe (2001)
Zn	Goethite	139.1	Experiment	Trivedi and Axe (2001)
Zn	Goethite	endothermic	Experiment	Rodda *et al.* (1993)
Zn	NiO	43.7	Experiment	Mahmood *et al.* (2011)
Zn	TiO2	24.55	Experiment	Li *et al.* (2008)
Zn	Hematite nanoparticles	–17.3	Experiment	Shipley *et al.* (2013)
Ni	HFO	58.2	Correlation model	Trivedi and Axe (2001)
Ni	Goethite	129.9	Experiment	Trivedi and Axe (2001)
Ni	HFO	62.2	Experiment	Trivedi and Axe (2001)
Ni	Natural iron oxide-coated sand	45.75	Experiment	Boujelben *et al.* (2009)
Hg	HFO	102.6	Correlation model	Trivedi and Axe (2001)
Hg	Polyacrylamide-grafted iron(III) oxide	26.34	Experiment	Manju *et al.* (2002)
Fe	HFO	83.7	Correlation model	Trivedi and Axe (2001)
Mn	HFO	88.0	Correlation model	Trivedi and Axe (2001)

Table S3.7 Tabulated sorption enthalpies for oxyanions by model optimization, corresponding LFER enthalpies for acid disassociation reactions (neutral and negative values. Last columns present enthalpies for relevant sorption reactions reported in the literature.

Pest optimization		Associated LFER reaction		Literature values by others	
Surface complex	**ΔH (kJ mol^{-1})**	**Disassociation reaction (Minteq database)**	**ΔH (kJ mol^{-1})**	**Experiment**	**ΔH (kJ mol^{-1})**
FeH_3SiO_4	-20 ± 5	$H_4SiO_4 = H_3SiO_4^- + H^+$	–20	Goethite sorption ΔH for CD music sorption model	–49.6[a]
FeH_2AsO_3	-41 ± 5	$H_2AsO_3^- + H^+ = H_3AsO_3$	–27.41	Goethite sorption ΔH for CD music sorption model	–26[b]
FeH_2BO_3	-23 ± 20	$H_3BO_3 = H_2BO_3^- + H^+$	–13	Sorption to Al-oxide	–13.6[c]
$FeCO_3^-$	-47 ± 10	$H^+ + CO_3^{-2} = HCO3^-$	–14.6		
$FeHCO_3$	-31 ± 8	$2H^+ + CO_3^{-2} = H_2CO_3$	–23.76		
$FeHPO4^-$	-61 ± 6	$2H^+ + PO_4^{-3} = H_2PO^{4-}$	–18		

Notes: sources of sorption enthalpies, [a]Kersten and Vlasova (2009b); [b]Kersten and Vlasova (2009a), [c]Goldberg *et al.* (1993).

S3.8 ADDITIONAL SIMULATION RESULTS OF THE AQUIFER THERMAL ENERGY STORAGE SCENARIO CASE STUDIES

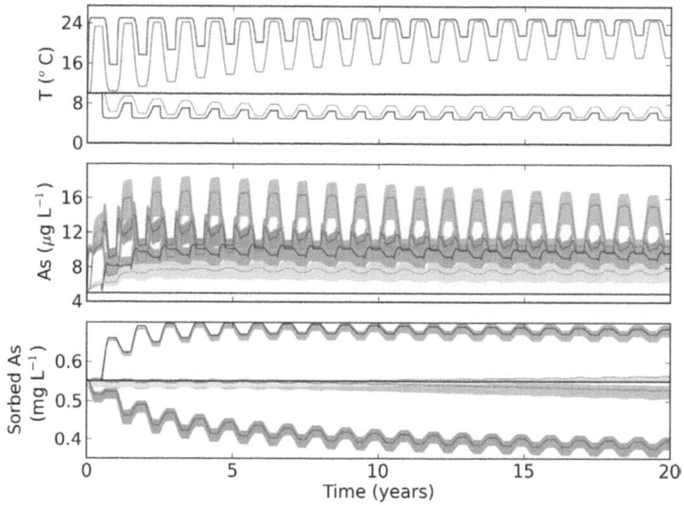

Figure S3.5 Simulation results for aqueous and sorbed arsenic concentration in virtual monitoring wells at 5 m and 15 m from the warm ATES well, shown in red and orange, respectively, and at 5 and 15 m from the cold ATES well, shown, in grey and light grey, respectively. The 95% predictive uncertainty ranges calculated with PEST[49] are shown in light shadings. Injection temperature in cold well is 5°C and in warm well at 25°C. Initial groundwater temperature is 11°C, and initial [As] = 5 µg L^{-1}.

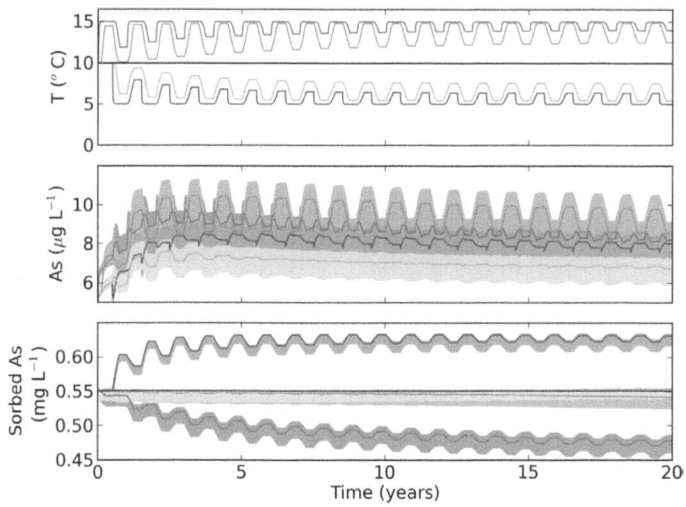

Figure S3.6 Simulation results for aqueous and sorbed arsenic concentration in virtual monitoring wells at 5 m and 15 m from the warm ATES well, shown in red and orange, respectively, and at 5 and 15 m from the cold ATES well, shown, in grey and light grey, respectively. The 95% predictive uncertainty ranges calculated with PEST[49] are shown in light shadings. Injection temperature in cold well is 7°C and in warm well at 15°C. Initial groundwater temperature is 11°C, and initial [As] = 5 µg L^{-1}.

IWA Publishing's authorised EU representative for General Product Safety Regulations is Diane D'Arras, 15 rue Duret, 75116 Paris, France, e-mail: safety@iwap.co.uk.

Printed and bound by CPI Group (UK) Ltd, Croydon, CR0 4YY
12/05/2026
02108866-0019